CARTESIAN TENSORS
AN INTRODUCTION

CARTESIAN TENSORS
AN INTRODUCTION

G. TEMPLE, F. R. S.
Sedleian Professor of Natural Philosophy
in the University of Oxford

DOVER PUBLICATIONS, INC.
Mineola, New York

Bibliographical Note

This Dover edition, first published in 2004, is an unabridged republication of the work published by Methuen & Co., Ltd., London, and John Wiley & Sons, Inc., New York, 1960.

Library of Congress Cataloging-in-Publication Data

Temple, George Frederick James, 1901-
Cartesian tensors : an introduction / G. Temple.
p. cm.
Includes index.
Originally published: London : Methuen ; New York : Wiley, c1960, in series: Methuen's monographs on physical subjects.
ISBN 0-486-43908-9 (pbk.)
1. Calculus of tensors. I. Title.

QA433.T4 2004
515'.63—dc22

2004049370

Manufactured in the United States of America
Dover Publications, Inc., 31 East 2nd Street, Mineola, N.Y. 11501

Contents

Contents

Preface

The purpose of this book is to provide an introduction to the theory of Cartesian tensors for first-year students pursuing an Honours course in Mathematics or Physics.

Tensor analysis was first forced upon the attention of theoretical physicists by the publication of Sir Arthur Eddington's 'Report on the Relativity Theory of Gravitation'.* In that report, however, tensor analysis was inevitably deployed on what was then the strange terrain of Riemannian geometry. In 1931 Sir Harold Jeffreys† had the happy idea of displaying tensor analysis on the familiar stage of three-dimensional Euclidean space. In this setting tensor analysis is freed from all irrelevant complications and is manifested in all its simplicity and power.

The excuse for writing another book on Cartesian tensors is that in the last thirty years the subject has been developed in a number of different directions which are of interest and importance to theoretical physicists.

The original definition of a tensor as a set of variables which are transformed cogrediently with the coordinate system with which they are associated has been replaced by the simpler and deeper definition now codified in the work of Bourbaki.‡ I have somewhat domesticated the native abstraction of this definition while preserving (I trust) its spirit and utility. Thus tensors are here defined as multilinear functions of direction, and this definition is found to simplify many theorems and to give a new unity to the subject.

The analysis of the structure of tensors (especially those of the

* *Phys. Soc. Lond.* 1915.

† *Cartesian Tensors*, Cambridge, 1931.

‡ N. Bourbaki, *Éléments de Mathématique, Les Structures Fondamentales de l'Analyse*, Livre II *Algèbre*, Chap. I. Structures Algèbriques.

second rank) in terms of spectral sets of projection operators is part of the very substance of quantum theory and therefore requires at least an elementary discussion.

The subject of isotropic tensors (whose components are the same in all orthogonal bases), always of fundamental importance in elasticity and hydrodynamics, has received new vigour from developments of the theory of group-representations and of abstract invariant theory.*

The development of spinor analysis, both as an algebraic discipline and as an integral part of quantum theory, appears to demand an introductory account, even in a work restricted to three-dimensional space.

I have therefore attempted to provide some initiation into these topics without quitting the confines of Euclidean space. I have, however, resisted the temptation to trespass too deeply into the territory of the hydrodynamicist and elastician, the quantum theorist and the relativist. For physicists the theory of tensors in a space–time manifold is so intimately associated with the special theory of relativity, and the theory of tensors in a Riemannian manifold with the general theory, that I have also held myself excused from these questions in this elementary introduction.

A number of examples have been devised to illustrate the general theory and to indicate certain extensions and applications.

For the pure mathematician this book can scarcely do more than encourage the study of linear algebras and of Riemannian geometry; for the applied mathematician and physicist it may foster an acquaintance with the theory and practice of that most useful language – Cartesian Tensors.

I am grateful to my colleague Professor C. A. Coulson who has kindly read this book in manuscript and has made many helpful suggestions.

G. T.

* H. Weyl, *The Classical Groups*, Princeton, 1931.

CHAPTER I

Vectors, Bases and Orthogonal Transformations

1.1 Introduction

Mathematicians and physicists have thoroughly exploited the day-dream in which M. des Cartes first envisaged the tri-rectangular frame of reference which we have named in his honour. But a Cartesian coordinate system is generally an external frame of reference with no organic or intrinsic relation to the mathematical or physical entities to be described. It is indispensable but irrelevant. The best we can do is express our mathematical descriptions in such a way that they preserve the same *form* in *any* Cartesian frame. Such a technique makes the frame harmless although necessary.

We need therefore some mathematical structures which shall be invariant under transformations from one frame of reference to another. Such structures are tensor algebra and analysis. But before studying these structures we need to study the nature of the transformations from one frame to another. We need not worry about such a triviality as a mere change of origin, but rotational transformations deserve serious consideration. We therefore begin with rotations and the vectors which they rotate.

1.2 The geometrical theory of vectors

In the geometrical theory of vectors a vector is defined, in effect, as an ordered pair of points, say $\overrightarrow{AB}$, although the harsh abstraction of this definition is often mitigated by the gloss that the vector is represented by an 'arrow', i.e. by the directed segment of the straight line AB.

It is at once evident that a vector so defined possesses both magnitude and direction, for the magnitude of the vector AB is naturally taken to be the distance $|AB|$ between A and B, and, if the points A, B, C, D (in this order) form the vertices of a parallelogram, then the vectors $\overrightarrow{AB}$ and $\overrightarrow{DC}$ are naturally described as equal in magnitude and similar in direction. But whereas the numerical representation

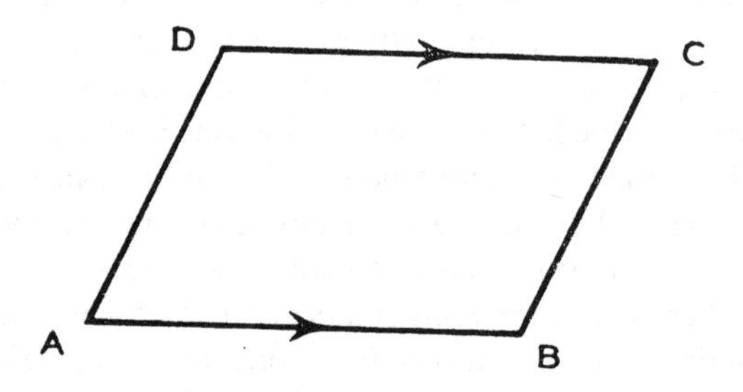

of the magnitude of a vector is obvious and direct, the numerical representation of its direction presents a problem.

The key to the solution of this problem is the fact that just as the position of a point can be specified only by reference to other points, so also the direction of a vector can be specified only by reference to other vectors. The relative direction of two vectors $\overrightarrow{OA}$ and $\overrightarrow{OB}$ is measured by the cosine of either of the angles, α or $2\pi - \alpha$, between the straight lines OA and OB. To give a complete description of the direction of a vector $\overrightarrow{OP}$ in three-dimensional space we need the cosines of the angles between the straight line OP and three other given straight lines OA, OB, OC, which are not coplanar.

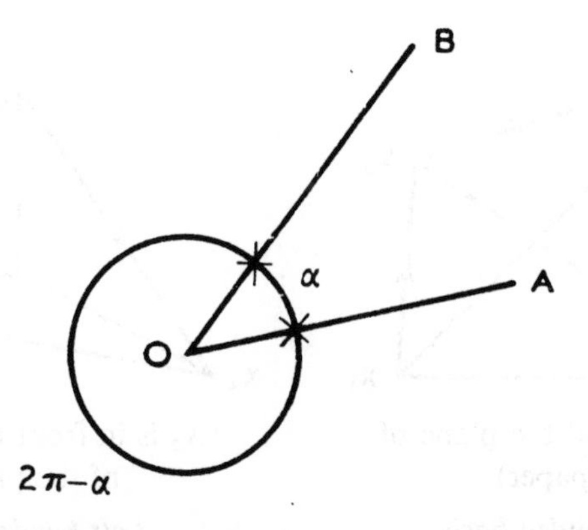

These same 'direction cosines' will then also specify the direction of any vector $\overrightarrow{QR}$ which is parallel to $\overrightarrow{OP}$.

1.3 Bases

The simplest way of systematizing the calculus of directions from a point P is to take three mutually perpendicular unit vectors $\overrightarrow{PX_1}$, $\overrightarrow{PX_2}$, $\overrightarrow{PX_3}$ as a *basis* of reference. There are then two species of bases – the left handed and the right handed. These are distinguished by drawing the triangle $X_1X_2X_3$ and viewing it from the point P. The vectors $\overrightarrow{X_2X_3}$; $\overrightarrow{X_3X_1}$; $\overrightarrow{X_1X_2}$ specify a sense of circulation around the triangle $X_1X_2X_3$. If this sense is right handed as seen from P, the basis is described as right handed; if the sense of circulation is left handed as seen from P, the basis is described as left handed. Right handedness and left handedness cannot be further analysed; they can only be demonstrated, as, for example, by reference to the directions in which the tendrils of the hop and the vine curl as they grow upwards.

Henceforward we shall employ only *right handed* bases, in accordance with what has become a standard and universal practice.

The direction of a vector $\overrightarrow{PQ}$, relative to the base $PX_1X_2X_3$, will then be specified by the ordered set of three direction cosines

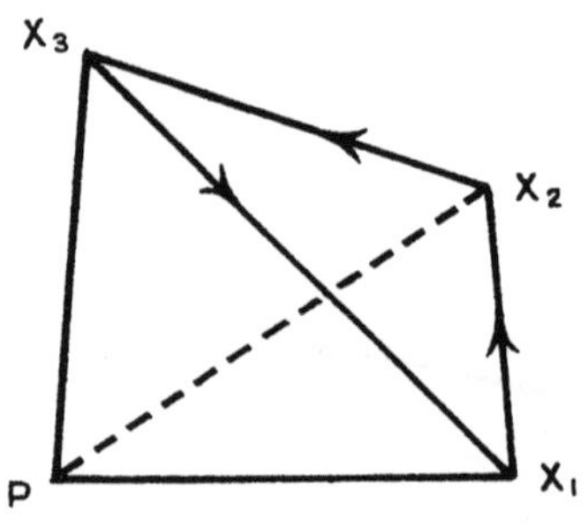

(X_2 is behind the plane of the paper)

Right handed basis

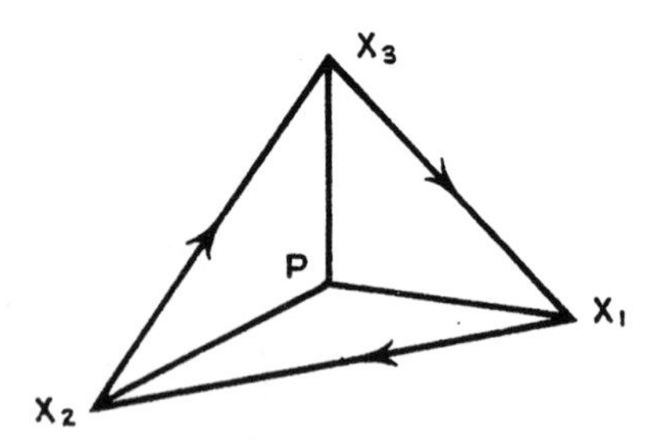

(X_2 is in front of the plane of the paper)

Left handed basis

(l_1, l_2, l_3), such that l_α = the cosine of either angle between $\overrightarrow{PX_\alpha}$ and $\overrightarrow{PQ}$.

Hence

$$-1 \leqslant l_\alpha \leqslant +1 \qquad (\alpha = 1, 2, 3).$$

Also, by taking the distance $|PQ|$ to be unity, it follows from Pythagoras' Theorem that

$$l_1^2 + l_2^2 + l_3^3 = 1. \tag{1.3.1}$$

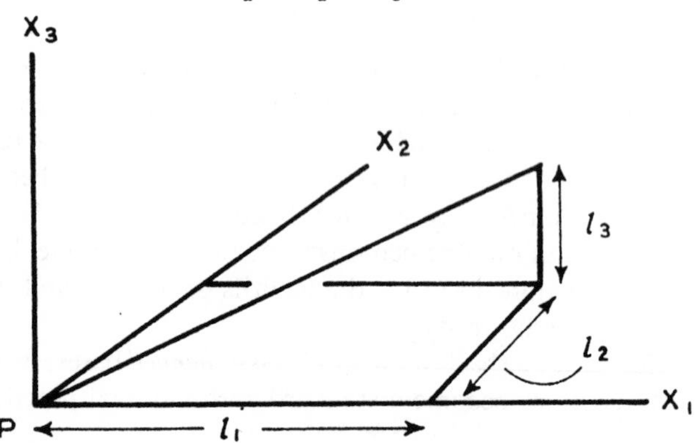

Now consider two vectors $\overrightarrow{PQ}$ and $\overrightarrow{PR}$ with direction cosines

(l_1, l_2, l_3) and (m_1, m_2, m_3). Then, by a well-known theorem of coordinate geometry in three dimensions, if θ is either of the angles between the straight lines PQ and PR,

$$\cos\theta = l_1m_1 + l_2m_2 + l_3m_3. \tag{1.3.2}$$

1.4 The summation convention

It is convenient to introduce at once a convention due to Einstein which considerably lightens the burden of writing, or setting up in type, such expressions as

$$l_1m_1 + l_2m_2 + l_3m_3 = \sum_{\alpha=1}^{3} l_\alpha m_\alpha.$$

This convention is that any expression, such as

$$l_\alpha m_\alpha,$$

in which a suffix, such as α, is repeated, is to be interpreted as the sum of all the values which $l_\alpha m_\alpha$ can take as α takes the values 1, 2, 3, i.e.

$$l_\alpha m_\alpha = l_1m_1 + l_2m_2 + l_3m_3.$$

Also any expression in which two or more suffixes $\alpha, \beta, \ldots$ are each repeated is to be interpreted as the sum of all the values which it can take as $\alpha, \beta, \ldots$ take the values 1, 2, 3, e.g.

$$A_{\alpha\beta}u_\alpha v_\beta = \sum_{\alpha,\beta} A_{\alpha\beta}u_\alpha v_\beta, \quad B_{\alpha\beta\gamma}u_\alpha v_\beta w_\gamma = \sum_{\alpha,\beta,\gamma} B_{\alpha\beta\gamma}u_\alpha v_\beta w_\gamma.$$

With this 'summation convention',

$$l_\alpha m_\alpha \equiv l_\beta m_\beta,$$

so that the identity of the letter used for a repeated suffix has no significance. Hence the repeated suffixes are called 'dummy suffixes'.

Similarly any symbol in brackets with a literal suffix, such as

$$(l_\alpha),$$

is taken to mean the ordered set (l_1, l_2, l_3).

1.5 The components of a vector

If the vector $\overrightarrow{PQ}$ has a magnitude

$$u = |PQ|,$$

and if its direction cosines relative to a base $PX_1X_2X_3$ are (l_α), then the *components* of the vector $\overrightarrow{PQ}$ in the directions (PX_α) are defined to be the numbers

$$u_\alpha = ul_\alpha,$$

(which may be positive or negative). These components are the lengths of the orthogonal projections of the vector $\overrightarrow{PQ}$ on the directions (PX_α).

The *component* of the vector $\overrightarrow{PQ}$ in the direction of any other vector $\overrightarrow{PR}$ with direction cosines (m_α) is similarly defined to be

$$u \cos \theta,$$

where θ is either of the angles between the lines PQ and PR. Now by (1.3.2),

$$\begin{aligned} u \cos \theta &= u(l_\alpha m_\alpha) \\ &= (ul_\alpha)m_\alpha \\ &= u_\alpha m_\alpha. \end{aligned}$$

Hence *the component of the vector* $\overrightarrow{PQ}$ *in the direction* (m_α) *is a linear function of the direction cosines* (m_α).

Example: Any vector u_α can be expressed in the form

$$u_\alpha = (u_\beta l_\beta)l_\alpha + (u_\beta m_\beta)m_\alpha + (u_\beta n_\beta)n_\alpha,$$

where the directions (l_α), (m_α), (n_α) form a base.

This elementary theorem enables us to give a rigorous interpretation to the familiar description of a vector as a 'directed magnitude'. Just as an ordinary (scalar) function $f(x)$ is completely specified by its numerical value for each assigned value of the argument x, so a vector $\overrightarrow{PQ}$ is completely specified by the numerical value of its component for each assigned direction (m_α). With any vector u there is therefore associated a function of direction, say $u(m_1, m_2, m_3)$, which determines its component in the direction (m_α). This by itself is not sufficient to characterize a vector, for the same statement is true of 'pseudo-vectors' such as moments of inertia. What really characterizes a true vector is that the associated function of direction $u(m_\alpha)$ is real, single-valued and *linear* in the direction cosines (m_α), so that

$$u = u(m_\alpha) = u_\alpha m_\alpha. \tag{1.5.1}$$

The advantages of associating a vector u with the linear function of direction, $u(m_\alpha) = u_\alpha m_\alpha$, as will appear later, are that the 'transformation laws' appear as an immediate deduction, and that the whole theory of tensors can be developed as a theory of *multilinear* functions of several directions.

1.6 Transformations of base

Since the direction and components of a vector are specified by reference to an arbitrary base, it is necessary to determine the changes produced in the direction and components by a change of base.

We therefore consider two bases (PX_α) and (PY_α). The base (PY_α) is completely specified in terms of the base (PX_α) by the nine numbers $T_{\alpha\beta}$, where

$$T_{\alpha\beta} = \text{the cosine of either angle between } PY_\alpha \text{ and } PX_\beta.$$

These numbers can be arranged in a square matrix, in which $T_{\alpha\beta}$ is the element in row α and column β.

It will be obvious that these nine direction cosines are far from being independent, and furthermore that, in general,

$$T_{\alpha\beta} \neq T_{\beta\alpha}.$$

Similarly the base (PX_α) is completely specified in terms of the base (PY_α) by the nine numbers $T'_{\alpha\beta}$, where $T'_{\alpha\beta}$ = the cosine of the angle between PX_α and PY_β.

Obviously

$$T'_{\alpha\beta} = T_{\beta\alpha},$$

so that the matrix T', formed with the elements $T'_{\alpha\beta}$, is the transpose of the matrix T, formed with the elements $T_{\alpha\beta}$.

Now let the direction of a vector $\overrightarrow{PQ}$ be specified by the direction cosines (l_α) relative to the base (PX_α) and by the direction cosines

(m_α) relative to the base (PY_α). Then

m_α = the projection of the line PQ on the direction PY_α
= the sum of the projections on PY_α of the projections of PQ in the directions PX_1, PX_2, PX_3
$= T_{\alpha 1}l_1 + T_{\alpha 2}l_2 + T_{\alpha 3}l_3$,

by the preceding definitions. Hence

$$m_\alpha = T_{\alpha\beta}l_\beta. \qquad (1.6.1)$$

Similarly
$$l_\alpha = T'_{\alpha\beta}m_\beta = m_\beta T_{\beta\alpha}. \qquad (1.6.2)$$

These formulae express the relation between directions relative to the two bases (PX_α) and (PY_α).

1.7 Properties of the transformation matrix T

The nine direction cosines $T_{\alpha\beta}$ are in fact connected by six independent relations which are easily derived as follows:

By using different dummy suffixes we can express $m_\alpha m_\alpha$ or $m_1^2 + m_2^2 + m_3^2$ in the form

$$m_\alpha m_\alpha = (T_{\alpha\beta}l_\beta)(T_{\alpha\gamma}l_\gamma) = (T_{\alpha\beta}T_{\alpha\gamma})(l_\beta l_\gamma).$$

But
$$m_\alpha m_\alpha = 1 \quad \text{and} \quad l_\alpha l_\alpha = 1.$$

Hence
$$(T_{\alpha\beta}T_{\alpha\gamma})(l_\beta l_\gamma) = l_\beta l_\beta,$$

for *any* set of direction cosines (l_β).

If we equate the coefficients of $l_\beta l_\gamma$ on each side of this identity, we find that

$$T_{\alpha\beta}T_{\alpha\gamma} = 0 \text{ if } \beta \neq \gamma \quad \text{or } 1 \text{ if } \beta = \gamma \qquad (1.7.1)$$

(no summation with respect to β being implied in the latter case).

Similarly we find that

$$l_\alpha l_\alpha = (T'_{\alpha\beta}T'_{\alpha\gamma})m_\beta m_\gamma = m_\beta m_\beta.$$

Hence
$$T'_{\alpha\beta}T'_{\alpha\gamma} = 0 \text{ if } \beta \neq \gamma \quad \text{or } 1 \text{ if } \beta = \gamma. \qquad (1.7.2)$$

These properties (1.7.1) and (1.7.2) of the transformation matrix T can be written and remembered more easily by introducing the unit matrix U, whose elements are the Knonecker symbols $\delta_{\alpha\beta}$, defined as

$$\delta_{\alpha\beta} = 0 \text{ if } \alpha \neq \beta \quad \text{or } 1 \text{ if } \alpha = \beta. \qquad (1.7.3)$$

Thus

$$U = \begin{Vmatrix} \delta_{11} & \delta_{12} & \delta_{13} \\ \delta_{21} & \delta_{22} & \delta_{23} \\ \delta_{31} & \delta_{32} & \delta_{33} \end{Vmatrix} = \begin{Vmatrix} 1 & 0 & 0 \\ 0 & 1 & 0 \\ 0 & 0 & 1 \end{Vmatrix}$$

It is also advantageous to write products of matrices so that the dummy suffixes are adjacent. Thus (1.7.1) and (1.7.2) can be written as

$$\begin{aligned} T'_{\beta\alpha}T_{\alpha\gamma} &= \delta_{\beta\gamma} \\ T_{\beta\alpha}T'_{\alpha\gamma} &= \delta_{\beta\gamma} \end{aligned} \qquad (1.7.4)$$

or even more compactly still as

$$\begin{aligned} T'T &= U \\ TT' &= U. \end{aligned} \qquad (1.7.5)$$

The corresponding transformation law for direction cosines may also be abbreviated to

$$\begin{aligned} m &= Tl, \quad l = T'm, \\ &= lT,' \quad\;\; = mT. \end{aligned} \qquad (1.7.6)$$

The transformation matrix T is also subject to a further condition when we restrict ourselves to transformations from one right handed base (PX_α) to another (PY_α), viz. that the determinant of the matrix T, det T, is equal to $+1$. All we can note here is that, by the theory of determinants, it follows from (1.7.4) or (1.7.5) that

$$(\det T').(\det T) = 1.$$

But $$(\det T') = (\det T).$$

Hence $$(\det T)^2 = 1,$$

and $$\det T = \pm 1. \qquad (1.7.7)$$

In fact it can be shown (see example 3, p. 13) that det $T = 1$ if (PX_α) and $(PY)_\alpha$ are both right handed or both left handed, but that det $T = -1$ if one is right handed and the other is left handed.

1.8 The orthogonal group

In this section we will not restrict ourselves to right handed bases. It will still be true, however, that, under a transformation from a basis PX_α to another basis PY_α, any vector remains unchanged in length. If (u_α) and (v_α) are the components of the same vector PQ in these two bases, and if S is the transformation matrix, then

$$v_\beta = S_{\beta\alpha}u_\alpha$$

and
$$v_\beta v_\beta = u_\alpha u_\alpha.$$

Such length-preserving transformations are called 'orthogonal transformations'.

Now let us carry out a second transformation from the basis PY_α to the basis PZ_α. Let the transformation matrix be T and let (w_α) be the components of the vector PQ in the third basis PZ_α. Then

$$w_\gamma = T_{\gamma\beta}v_\beta,$$

and
$$w_\gamma w_\gamma = v_\beta v_\beta.$$

It is clear that the direct transformation from PX_α to PZ_α must also be an orthogonal transformation, since

$$w_\gamma w_\gamma = u_\alpha u_\alpha.$$

In this transformation

$$w_\gamma = T_{\gamma\beta}S_{\beta\alpha}u_\alpha = U_{\gamma\alpha}u_\alpha, \text{ say.}$$

Hence the matrix of this transformation, U, is the product of the matrix S by the matrix T.

Any two orthogonal transformations, S and T, therefore possess a 'product' TS. The product TS is in general different from the product ST, so that multiplication is *not* commutative.

We can also invert any orthogonal transformation, for

$$u_\alpha = S'_{\alpha\beta}v_\beta,$$

where S' is the transpose of S. Hence any orthogonal transformation S possesses an inverse S^{-1}, which *for such transformations* is merely the transpose S' of S.

Now any set of transformations T_n which includes the inverse $T_n{}^{-1}$ of any transformation T_n, and the product T_kT_j of any two transformations T_j, T_k is called a *group* of transformations. Hence the orthogonal transformations form a group, called the 'orthogonal group'.

We have already proved (1.7.7) that, if T is the matrix of any orthogonal transformation, then

$$\det T = \pm 1.$$

The orthogonal transformations therefore fall into two classes, (1) the 'proper' orthogonal transformations, for which

$$\det T = +1,$$

and (2) the 'improper' orthogonal transformations, for which

$$\det T = -1.$$

The proper orthogonal transformations themselves form a group, for if

$$\det S = 1 \quad \text{and} \quad \det T = 1,$$

then
$$\det(TS) = (\det T).(\det S) = 1.$$

This group is a subgroup of the complete orthogonal group. The improper orthogonal transformations do not form a group.

1.9 Examples

(1) A rotation about PX_3 through the angle α has the transformation matrix

$$T = \begin{Vmatrix} \cos\alpha & \sin\alpha & 0 \\ -\sin\alpha & \cos\alpha & 0 \\ 0 & 0 & 1 \end{Vmatrix}$$

and $\det T = +1$.

(2) A reflexion in the plane $x_2 \cos\alpha = x_1 \sin\alpha$ has the transformation matrix

$$T = \begin{Vmatrix} \cos 2\alpha & \sin 2\alpha & 0 \\ \sin 2\alpha & -\cos 2\alpha & 0 \\ 0 & 0 & 1 \end{Vmatrix}$$

with determinant $\det T = -1$.

(3) The transformation matrix for a rotation is most easily obtained by the use of vector algebra.

Let the rotation be through an angle ω about an axis $P\Omega$ *of unit length* and direction cosines (λ_α). Let the point A with coordinates (l_α) be carried to the point B with coordinates (m_α). Let N be the foot of the perpendiculars from A and B on to $P\Omega$.

Draw AC in the plane of ANB and perpendicular to NA to meet NB produced in C. Then

$$\overrightarrow{PB} = (1 - \cos\omega)\overrightarrow{PN} + \cos\omega.\overrightarrow{PC},$$
$$\overrightarrow{PN} = \overrightarrow{P\Omega}(\overrightarrow{PA}.\overrightarrow{P\Omega}),$$
$$\overrightarrow{PC} = \overrightarrow{PA} + \tan\omega.\overrightarrow{P\Omega}\wedge\overrightarrow{PA}.$$

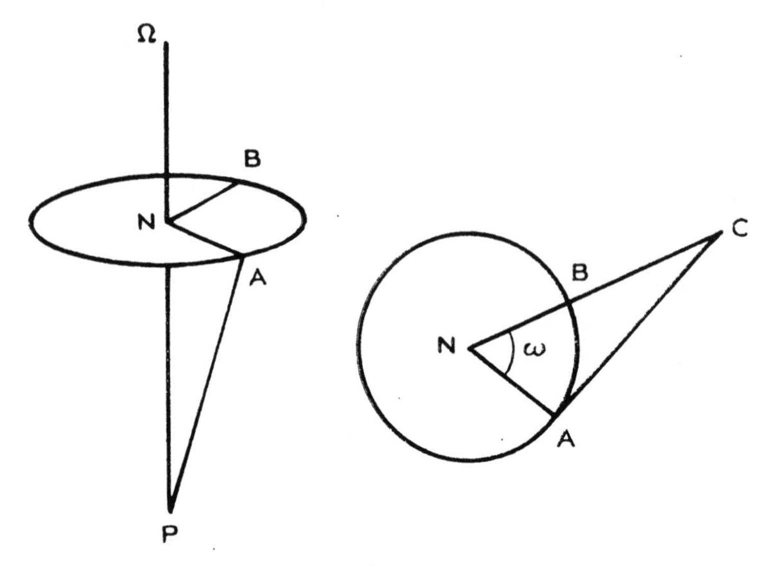

Hence $m_1 = (1 - \cos\omega)(l_\alpha\lambda_\alpha)\lambda_1 + \cos\omega.l_1 + \sin\omega.(\lambda_2 l_3 - \lambda_3 l_2)$, with two similar equations, i.e.

$$T_{11} = \cos\omega + (1 - \cos\omega)\lambda_1^2,$$
$$T_{23} = (1 - \cos\omega)\lambda_2\lambda_3 - \sin\omega.\lambda_1,$$
$$T_{32} = (1 - \cos\omega)\lambda_3\lambda_2 + \sin\omega.\lambda_1, \text{ etc.}$$

The determinant of T is a continuous function of ω, which reduces to $+1$ as $\omega \longrightarrow 0$. Hence det T is always equal to $+1$.

(4) A reflexion in the plane $\lambda_\alpha x_\alpha = 0$, whose normal has direction cosines (λ_α), has the transformation matrix

$$T_{\alpha\beta} = \delta_{\alpha\beta} - 2\lambda_\alpha\lambda_\beta,$$

and $$\det T = -1.$$

(5) A 'half-turn', i.e. a rotation through 180° about an axis with direction cosines (λ_α) has the transformation matrix

$$T_{\alpha\beta} = -\,\delta_{\alpha\beta} + 2\lambda_\alpha\lambda_\beta,$$

with
$$\det T = +\,1.$$

(6) A reflexion in the origin has the transformation matrix $T_{\alpha\beta} = -\,\delta_{\alpha\beta}$; and the product of three reflexions in three mutually orthogonal planes is a reflexion in their common point of intersection.

(7) Show that in the matrix T of any *proper* orthogonal transformation, any element $T_{\alpha\beta}$ is equal to its cofactor. Hence show that

$$\det\,(T_{\alpha\beta} - \delta_{\alpha\beta}) = 0.$$

(8) Deduce from the preceding result that there exists a vector λ_α such that

$$T_{\alpha\beta}\lambda_\beta = \lambda_\alpha,$$

in any proper orthogonal transformation. (The vector λ_α lies along the axis of the rotation specified by T.)

(9) By taking a basis with PY_3 along this vector (λ_α), show that any proper orthogonal transformation is a rotation, and therefore preserves the chirality of any base (i.e. a right handed base is transformed into a right handed base, etc.).

(10) By considering reflexions in the planes

$$x_2 \cos\alpha = x_1 \sin\alpha, \quad x_2 \cos\beta = x_1 \sin\beta,$$

show that any rotation is the product of two reflexions.

CHAPTER II

The Definition of a Tensor

2.1 Introduction

The purpose of this chapter is to introduce the algebraical definition of a *tensor* as a multilinear function of direction. This definition is the simplest form of the abstract definition of a tensor adopted by Bourbaki and it gives a new unity to the whole subject of tensor algebra and analysis by suggesting simple and direct proofs of many fundamental theorems.

We commence with some specific simple examples of multilinear functions of direction before giving the formal definition of a tensor.

2.2 Geometrical examples of multilinear functions of direction

Two elementary formulae of coordinate geometry provide simple examples of multilinear functions of direction of ranks two and three, i.e. functions of two or three directions respectively. (A vector will be regarded as a tensor of rank one.)

The cosine of either angle between the two directions (l_α) and (m_α) is

$$\cos\theta = l_\alpha m_\alpha.$$

If the right hand side of this equation is written as

$$\delta_{\alpha\beta} l_\alpha m_\beta,$$

where $\delta_{\alpha\beta}$ is the Knonecker symbol introduced in equation 1.7.3, it is evident that $\cos\theta$ is a bilinear function of (l_α) and (m_α) with coefficients $\delta_{\alpha\beta}$. We can therefore regard the numbers $\delta_{\alpha\beta}$ as the components of a tensor U of the second rank. Since the numbers $\delta_{\alpha\beta}$ are the elements of the unit matrix in three dimensions, U is called the 'unit tensor'.

The volume of the parallelepiped with unit edges in the directions (l_α), (m_α) and (n_α) is

$$V = \begin{vmatrix} l_1 & l_2 & l_3 \\ m_1 & m_2 & m_3 \\ n_1 & n_2 & n_3 \end{vmatrix} \qquad (2.2.1)$$

if the directions (l_α), (m_α), (n_α) form a right handed triad. Now we can write this determinant in the form

$$V = \varepsilon_{\alpha\beta\gamma} l_\alpha m_\beta n_\gamma,$$

where the symbol $\varepsilon_{\alpha\beta\gamma}$ is defined as follows:

$$\varepsilon_{123} = \varepsilon_{231} = \varepsilon_{312} = +1,$$
$$\varepsilon_{321} = \varepsilon_{213} = \varepsilon_{132} = -1,$$

and $\varepsilon_{\alpha\beta\gamma} = 0$ if any two suffixes are the same.

In other words

$$\varepsilon_{\alpha\beta\gamma} = \begin{array}{l} +1, \text{ if } (\alpha, \beta, \gamma) \text{ is an even permutation of } (1, 2, 3), \\ -1, \text{ if } (\alpha, \beta, \gamma) \text{ is an odd permutation of } (1, 2, 3), \\ \ \ 0, \text{ if } (\alpha, \beta, \gamma) \text{ has any other set of values.} \end{array}$$

It is clear that the volume V is a trilinear function of (l_α), (m_α) and (n_α) with coefficients $\varepsilon_{\alpha\beta\gamma}$. We can therefore regard the numbers $\varepsilon_{\alpha\beta\gamma}$ as the components of a tensor A of the third rank. A is called the 'alternating tensor'.

The tensors U and A possess the remarkable property that their components $\delta_{\alpha\beta}$ and $\varepsilon_{\alpha\beta\gamma}$ have the same numerical values for all bases. They are in fact the only tensors of the second and third rank respectively with this property as will be proved later (Chap. VI).

These symbols, $\delta_{\alpha\beta}$ and $\varepsilon_{\alpha\beta\gamma}$, are of considerable importance in tensor theory. Familiarity with their properties can be gained by verifying the following identities of which the most important is number (2).

Exercises: (1) If $(m_\alpha n_\alpha) = \cos\theta$, $(n_\alpha l_\alpha) = \cos\phi$, $(l_\alpha m_\alpha) = \cos\psi$, then the volume

$$V = 1 - \cos^2\theta - \cos^2\phi - \cos^2\psi + 2\cos\theta\cos\phi\cos\psi.$$

(2) $$\sum_{\gamma} \varepsilon_{\alpha\beta\gamma} . \varepsilon_{\rho\sigma\gamma} = \delta_{\alpha\rho}\delta_{\beta\sigma} - \delta_{\alpha\sigma}\,\delta_{\beta\rho}.$$

(This is the tensor equivalent of the vector identity,

$$(\mathbf{l}_{\wedge}\,\mathbf{m}).(\boldsymbol{\lambda}_{\wedge}\,\boldsymbol{\mu}) = (\mathbf{l}.\boldsymbol{\lambda})(\mathbf{m}.\boldsymbol{\mu}) - (\mathbf{m}.\boldsymbol{\lambda})(\mathbf{l}.\boldsymbol{\mu})\ )$$

(3) $$\varepsilon_{\alpha\beta\gamma}u_{\alpha a}u_{\beta b}u_{\gamma c} = \varepsilon_{abc} \det u$$
$$\varepsilon_{abc}u_{\alpha a}u_{\beta b}u_{\gamma c} = \varepsilon_{\alpha\beta\gamma} \det u$$

2.3 Examples of multilinear functions of direction in rigid dynamics

It is almost trivial to observe that displacements, velocities, accelerations and forces are instances of vectors, i.e. of tensors of the first rank. Two important dynamical tensors of the second and third ranks respectively are the inertia tensor and the tensor which gives the moment of a force about a skew axis.

We consider a distribution of particles with a typical particle of mass m at the point (x_α). The moment of inertia about an axis PQ (l_α) is, in the notation of the figure

$$\begin{aligned}\Sigma\, m(r \sin \theta)^2 &= \Sigma\, mr^2 - \Sigma\, m(r \cos \theta)^2 \\ &= \Sigma\, m(x_\alpha x_\alpha) - \Sigma\, m(x_\alpha l_\alpha)^2.\end{aligned}$$

The product of inertia about two perpendicular axis PQ and PR, (l_α) and (λ_α) is

$$\begin{aligned}\Sigma\, m(r \cos \theta)(r \cos \phi) &= \Sigma\, m(x_\alpha l_\alpha)(x_\alpha \lambda_\alpha) \\ &= -\,\Sigma\, m(x_\alpha x_\alpha)(l_\alpha \lambda_\alpha) + \Sigma\, m(x_\alpha l_\alpha)(x_\alpha \lambda_\alpha)\end{aligned}$$

since $(l_\alpha \lambda_\alpha) = 0$.

It appears therefore that the moment of inertia can be regarded as the product of inertia about two coincident axes, with the sign reversed. It is therefore natural to introduce the 'inertia tensor' I defined as a bilinear function of two directions (l_α) and (λ_α) by the formula

$$I(l, \lambda) = \Sigma m(x_\beta x_\beta)(l_\alpha \lambda_\alpha) - \Sigma m(x_\alpha l_\alpha)(x_\beta \lambda_\beta).$$

In the usual notation the components of the inertia tensor are given by

$$\begin{aligned}I(l, \lambda) = Al_1\lambda_1 + Bl_2\lambda_2 + Cl_3\lambda_3 &- F(l_2\lambda_3 + l_3\lambda_2) \\ &- G(l_3\lambda_1 + l_1\lambda_3) - H(l_1\lambda_2 + l_2\lambda_1),\end{aligned}$$

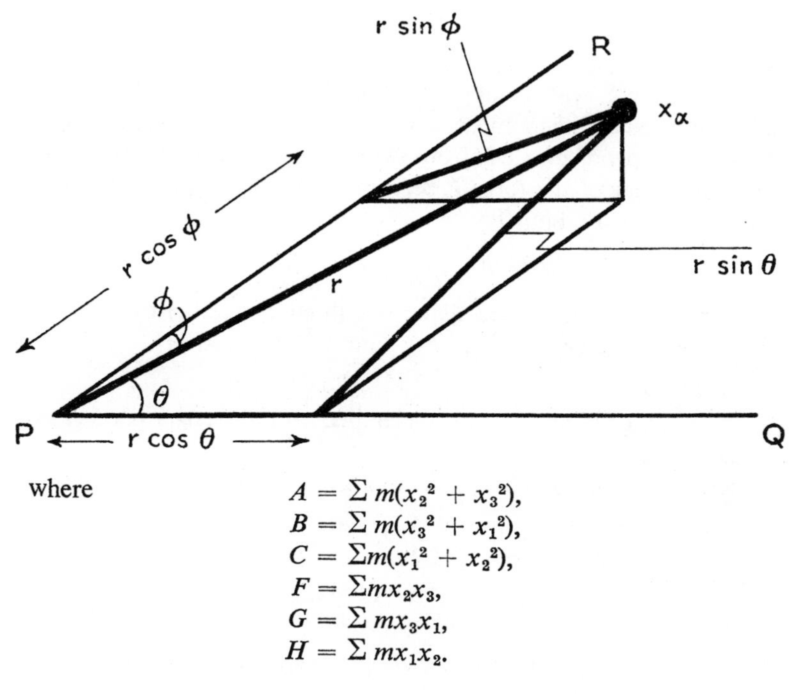

where

$$A = \Sigma\, m(x_2^2 + x_3^2),$$
$$B = \Sigma\, m(x_3^2 + x_1^2),$$
$$C = \Sigma m(x_1^2 + x_2^2),$$
$$F = \Sigma m x_2 x_3,$$
$$G = \Sigma\, m x_3 x_1,$$
$$H = \Sigma\, m x_1 x_2.$$

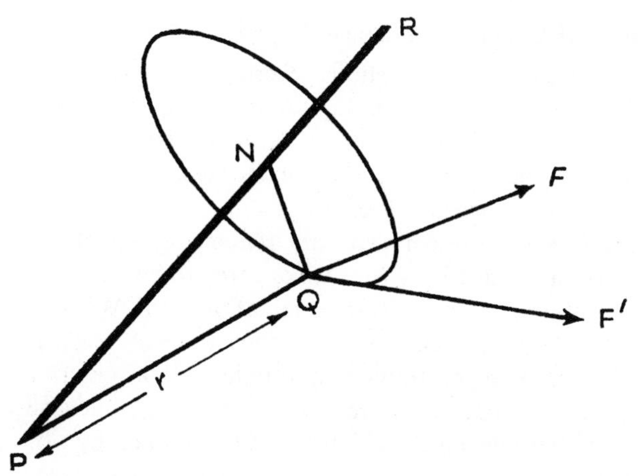

Next we calculate the moment of a force about an arbitrary axis.

Let a force **F** with components $(m_\alpha F)$ act at a point Q (x_α) or (rl_α). To calculate its moment about an axis PR in the direction (λ_α), draw the perpendicular QN from Q on to the axis PR, and let F' be the component of the force **F** perpendicular to PN and PQ. Then the moment of **F** about PR is

$$NQ.F' = PQ.\sin QPN.F'$$

= the volume of a parallelepiped with edges parallel to $\overrightarrow{PN}$, $\overrightarrow{PQ}$ and **F** of lengths unity, $|PQ|$ and $|F|$ respectively.

$= rFV,$

where V is the volume of a parallelepiped with unit edges parallel to PN, PQ and **F**.

Hence by equation (2.2.1), the moment of the force about PR is

$$M(\lambda, l, m) = rF \begin{vmatrix} \lambda_1 & \lambda_2 & \lambda_3 \\ l_1 & l_2 & l_3 \\ m_1 & m_2 & m_3 \end{vmatrix}$$

$$= rF\varepsilon_{\alpha\beta\gamma}\lambda_\alpha l_\beta m_\gamma.$$

This is manifestly a trilinear function of (λ_α), (l_α) and (m_α).

2.4 The stress tensor in continuum dynamics

In analysing the forces which are exerted on a part R of a continuous medium, be it solid, liquid or gas, by the surrounding medium, we distinguish between the 'body forces' which act directly on the volume elements of R and the 'surface forces' which act on the elements of the surface S of R.

We are free to consider arbitrary regions R, and if we fix our attention on a specified point P, we are free to consider surfaces S passing through P with arbitrary orientations at P. We are thus led to consider the force acting on a surface element at P. The component of this force per unit area at a point P in a prescribed direction (l_α) will depend, in general, not only on (l_α) but also on the direction of the normal to S (λ_α). Let it be denoted by $f(l, \lambda)$.

The action and reaction transmitted across any surface element must be equal and opposite (assuming that the continuous medium is subject to Newton's third law of motion). Hence

$$f(l, \lambda) = -f(l, -\lambda).$$

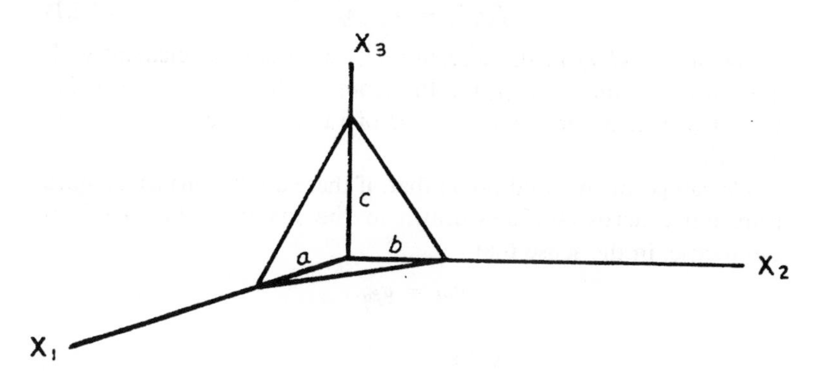

We shall prove that $f(l, \lambda)$ is a bilinear function of (l_α) and of (λ_α), by considering the equilibrium of the matter inside the small tetrahedron with vertices at the points $(0, 0, 0)$, $(a, 0, 0)$, $(0, b, 0)$, $(0, 0, c)$, where a, b, c are small quantities, each say of order η.

Let $\sigma_{\alpha\beta}$ denote the stress at P in the direction PX_α on the surface element with inward normal in the direction PX_β.

Then the following table gives the areas and stresses, parallel to PX_α, on the four faces of the tetrahedron, correct to terms of order η.

Face	*Area*	*Direction cosines of inward drawn normal*	*Stress*
$x_1 = 0$	$\frac{1}{2}bc$	1, 0, 0	$l_\alpha\sigma_{\alpha 1}$
$x_2 = 0$	$\frac{1}{2}ca$	0, 1, 0	$l_\alpha\sigma_{\alpha 2}$
$x_3 = 0$	$\frac{1}{2}ab$	0, 0, 1	$l_\alpha\sigma_{\alpha 3}$
$\frac{x_1}{a} + \frac{x_2}{b} + \frac{x_3}{c} = 1$	A	$-\lambda_1, -\lambda_2, -\lambda_3$	$f(l, -\lambda)$

It follows that the resultant, parallel to PX_α, of all the surface forces is

$$\tfrac{1}{2}bcl_\alpha\sigma_{\alpha 1} + \tfrac{1}{2}cal_\alpha\sigma_{\alpha 2} + \tfrac{1}{2}abl_\alpha\lambda_{\alpha 3} + Af(l, -\lambda) + 0(\eta^3)$$
$$= A\{\lambda_1 l_\alpha\sigma_{\alpha 1} + \lambda_2 l_\alpha\sigma_{\alpha 2} + \lambda_3 l_\alpha\sigma_{\alpha 3} - f(l, \lambda)\} + 0(\eta^3).$$

But the resultant of the body forces and inertial forces, being proportional to the mass (and therefore to the volume) of the tetrahedron, is $0(\eta^3)$. Therefore

$$f(l, \lambda) = \sigma_{\alpha\beta} l_\alpha \lambda_\beta. \tag{2.4.1}$$

The stress $f(l, \lambda)$ in the direction (l_α) on a surface element with normal in the direction (λ_α) is therefore a bilinear function of (l_α) and of (λ_α). It is therefore a tensor of rank two, called the 'stress tensor'.

We can go further and prove that, if there are no surface couples transmitted across surfaces drawn in the material, then $f(l, \lambda)$ is symmetric, in the sense that

$$\sigma_{\alpha\beta} = \sigma_{\beta\alpha}.$$

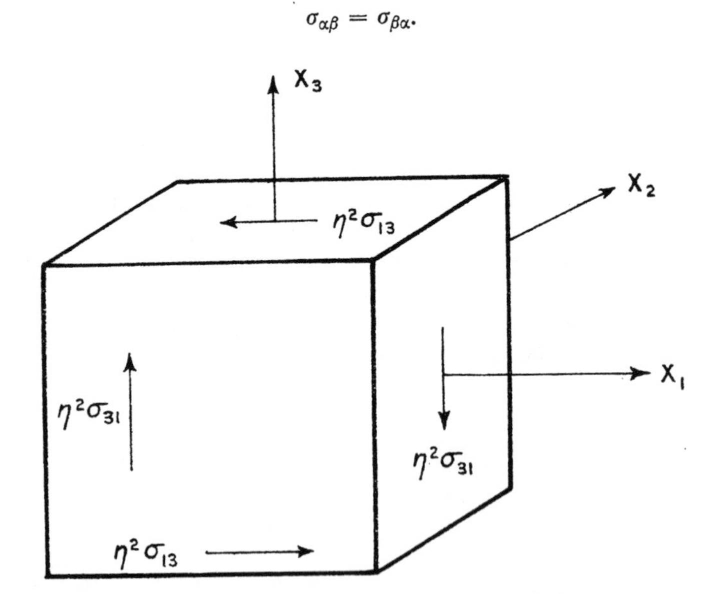

Consider the equilibrium of the cube of material bounded by the faces $x_1 = \pm \frac{1}{2}\eta$, $x_2 = \pm \frac{1}{2}\eta$, $x_3 = \pm \frac{1}{2}\eta$, and take moments about PX_2. Let $\sigma_{\alpha\beta}$ denote the stress tensor at the origin. The couples due to stresses on the faces $x_1 = \pm \frac{1}{2}\eta$ and $x_3 = \pm \frac{1}{2}\eta$ are

$$2\eta^2(\sigma_{31} - \sigma_{13}) + 0(\eta^3).$$

The couples due to stresses on the faces $x_2 = \pm \frac{1}{2}\eta$ are $0(\eta^3)$. The body couples (if any) and the inertial couple are $0(\eta^3)$ as before.

Hence $\qquad \sigma_{31} = \sigma_{13}.$

Similarly $\qquad \sigma_{12} = \sigma_{21}$

and $\qquad \sigma_{23} = \sigma_{32}.$

Therefore $\qquad \sigma_{\alpha\beta} = \sigma_{\beta\alpha}.$

A second rank tensor with components of this character is said to be symmetric.

The components $\sigma_{\alpha\beta}$ of the stress tensor in the base $PX_1X_2X_3$ form a symmetric matrix, of which the diagonal elements, σ_{11}, σ_{22}, σ_{33}, are called 'tensile' stresses and the off-diagonal elements, $\sigma_{23} = \sigma_{32}$, $\sigma_{31} = \sigma_{13}$. $\sigma_{12} = {}_{21}$, are called 'shearing' stresses.

Exercise: Show that if the mean tensile stress is zero, i.e. if $\sigma_{\alpha\alpha} = 0$, then there exist bases (actually infinite in number) in which the stress system consists of shearing stresses only.

For, if $\sigma_{\alpha\alpha} = 0$, then the cone C with equation $x_\alpha \sigma_{\alpha\beta} x_\beta = 0$ has sets of three perpendicular generators. If such a set $PY_1Y_2Y_3$ is taken as a base and if $\tau_{\alpha\beta}$ are the components of the stress tensor in this base, then the equation of C, viz. $y_\alpha \tau_{\alpha\beta} y_\beta = 0$, must reduce to

$$y_2\tau_{23}y_3 + y_3\tau_{31}y_1 + y_1\tau_{12}y_2 = 0$$

and τ_{11}, τ_{22}, τ_{33} must each be zero.

2.5 Formal definition of a tensor

In the light of these examples we can now give a formal definition of a tensor as follows:

A tensor is an invariant multilinear function of direction, i.e. it is a function of a number of directions which takes the form

$$S_{\alpha\beta\gamma\ldots} l_\alpha m_\beta n_\gamma \ldots \qquad (2.5.1)$$

in any particular basis, (l_α), (m_α), (n_α), . . . being the direction cosines of the directions in the basis, and which has the same numerical value in every basis. The number of directions involved is called the 'rank' of the tensor. A scalar quantity, such as density, can be regarded as a tensor of rank zero. Vectors, as we have seen in Chapter 1, can be regarded as tensors of rank unity.

The coefficients $S_{\alpha\beta\gamma\ldots}$ in the expression for a tensor in a prescribed basis are called the 'components' of the tensor in that basis. The relation between the coefficients $S_{\alpha\beta\gamma\ldots}$ and $T_{\alpha\beta\gamma\ldots}$ in two different bases PX_α and PY_α follows at once from the preceding definition.

Let us consider first the important case of a second rank tensor $S_{\alpha\beta}l_\alpha m_\beta$. In the shorthand notation introduced in § 1.7 the matrix of the components $S_{\alpha\beta}$ in the basis PX_α is denoted by S, and the direction cosines by l and m. In another basis PY_α, the matrix of the components $T_{\alpha\beta}$ is denoted by T and the direction cosines by λ and μ. The transformation equations for the direction cosines are (1.7.6)

$$\lambda = Rl, \quad \mu = Rm$$

or

$$l = R'\lambda = \lambda R, \quad m = R'\mu.$$

Now the expressions for the tensor in the bases PX_α and PY_β are respectively

$$lSm \equiv l_\alpha S_{\alpha\beta} m_\beta \quad \text{and} \quad \lambda T\mu \equiv \lambda_\alpha T_{\alpha\rho}\mu_\rho$$

Hence

$$\begin{aligned} \lambda T\mu &= lSm \\ &= \lambda R.S.R'\mu, \end{aligned}$$

or

$$\lambda_\alpha T_{\alpha\beta}\mu_\beta = \lambda_\alpha R_{\alpha\rho} S_{\rho\sigma} R'_{\sigma\beta}\mu_\beta.$$

Since this equation holds for all λ_α and μ_β, it follows that

$$T_{\alpha\beta} = R_{\alpha\rho} S_{\rho\sigma} R'_{\sigma\beta},$$

or

$$T = RSR'. \tag{2.5.2}$$

This is the transformation law connecting the components of a second rank tensor in two different bases. Unfortunately the shorthand matrix notation cannot be used for tensors of higher rank, but the corresponding transformation is easily found as follows:

$$\begin{aligned} T_{\alpha\beta\gamma\ldots}\lambda_\alpha\mu_\beta\nu_\gamma\ldots &= S_{\rho\sigma\tau\ldots} l_\rho m_\sigma n_\tau \ldots \\ &= S_{\rho\sigma\tau\ldots}\lambda_\alpha R_{\alpha\rho}\mu_\beta R_{\beta\sigma}\nu_\gamma R_{\gamma\tau} \ldots \end{aligned}$$

whence

$$T_{\alpha\beta\gamma\ldots} = R_{\alpha\rho} R_{\beta\sigma} R_{\gamma\tau} \ldots S_{\rho\sigma\tau\ldots} \tag{2.5.3}$$

Exercises: (1) It is well worth while to write out the transformation in full for a second rank tensor in two dimensional space and a rotation R in the plane PX_1X_2.

Then
$$\lambda_1 = l_1 \cos\alpha + l_2 \sin\alpha,$$
$$\lambda_2 = -\,l_1 \sin\alpha + l_2 \cos\alpha,$$

with similar equations for μ_1, μ_2 in terms of m_1, m_2.

Any tensor T of the second rank in two dimensions can be expressed in terms of the four fundamental tensors with the matrices

$$U = \begin{Vmatrix} 1 & 0 \\ 0 & 1 \end{Vmatrix} \qquad V = \begin{Vmatrix} 1 & 0 \\ 0 & -1 \end{Vmatrix}$$

$$W = \begin{Vmatrix} 0 & 1 \\ 1 & 0 \end{Vmatrix} \qquad Z = \begin{Vmatrix} 0 & 1 \\ -1 & 0 \end{Vmatrix}$$

as
$$T = \tfrac{1}{2}(T_{11} + T_{22})U + \tfrac{1}{2}(T_{11} - T_{22})V + \tfrac{1}{2}(T_{12} + T_{21})W + \tfrac{1}{2}(T_{12} - T_{21})Z.$$

Let these tensors become $\bar{U}$, $\bar{V}$, $\bar{W}$ and $\bar{Z}$ under the rotation R. Then

$$\begin{aligned} \lambda\bar{V}\mu &= lVm \\ &= l_1m_1 - l_2m_2 \\ &= \lambda V\mu \cos 2\alpha - \lambda W\mu \sin 2\alpha, \end{aligned}$$

i.e.
$$\bar{V} = V\cos 2\alpha - W\sin 2\alpha.$$

Similarly

$$\bar{W} = W\cos 2\alpha + V\sin 2\alpha,$$
$$\bar{U} = W \quad \text{and} \quad \bar{Z} = Z.$$

(2) The transformation law for the components of a tensor $T_{\alpha\beta}$ is exactly the same as for the product $u_\alpha v_\beta$, of the components of two vectors, u_α and v_α.

2.6 The angular velocity tensor

Consider a rigid body rotating about the fixed point P. Let a typical particle of the body move from the point with coordinates (X_α) at time $t = 0$ to the point (x_α) at time t, these coordinates being referred to the same fixed base at P. In a base carried with the rigid body the particle retains the same coordinates (X_α). Hence the relation between (x_α) and (X_α) must be of the form

$$x_\alpha = R_{\alpha\beta}X_\beta, \quad X_\beta = R'_{\beta\alpha}x_\alpha,$$

where R is a transformation matrix, which of course varies with the time t.

The velocity of the typical particle is therefore

$$\begin{aligned} \dot{x}_\alpha &= \dot{R}_{\alpha\beta} X_\beta \\ &= \dot{R}_{\alpha\beta} R_{\beta\gamma} x_\gamma \\ &= \Omega_{\alpha\gamma} x_\gamma, \end{aligned} \tag{2.6.1}$$

say, the dots indicating differentiation with respect to the time.

Now $\qquad RR' = U, \quad \text{and} \quad \dot{U} = O,$

so that $\qquad \Omega = \dot{R}R' = -R\dot{R}'$

and $\qquad \Omega' = -\dot{R}R' = -\Omega,$

i.e. $\qquad \Omega_{\gamma\alpha} = -\Omega_{\alpha\gamma}. \qquad (2.6.2)$

The matrix $\Omega_{\alpha\beta}$ is therefore antisymmetric in the suffixes α and β and has the form

$$\Omega = \begin{Vmatrix} 0 & -\omega_3 & +\omega_2 \\ \omega_3 & 0 & -\omega_1 \\ -\omega_2 & +\omega_1 & 0 \end{Vmatrix} \tag{2.6.3}$$

By (2.6.1) a particle at the point $x_\alpha = r\lambda_\alpha$ has a component velocity in the direction l_α equal to

$$l_\alpha \dot{x}_\alpha = r \,.\, l_\alpha \Omega_{\alpha\beta} \lambda_\beta,$$

whence the numbers $\Omega_{\alpha\beta}$ form the components of a tensor called the 'angular velocity tensor'. However

$$\begin{aligned} \dot{x}_1 &= \omega_2 x_3 - \omega_3 x_2, \\ \dot{x}_2 &= \omega_3 x_1 - \omega_1 x_3, \\ \dot{x}_3 &= \omega_1 x_2 - \omega_2 x_1, \end{aligned} \tag{2.6.4}$$

i.e. in vector notation, the velocity $\mathbf{v}$ is the vector product $\boldsymbol{\omega} \wedge \mathbf{x}$ of the vector $\boldsymbol{\omega}$ with components (ω_α) and the position vector $\mathbf{x} = (x_\alpha)$.

The vector character of $\boldsymbol{\omega}$ is more fully examined later in § 3.6.

CHAPTER III

The Algebra of Tensors

3.1 Introduction

The use of tensors in mathematical physics is greatly facilitated by a number of simple techniques, of which the most important are the methods of 'contraction' and differentiation. In this chapter we explain the method of contraction and the more important algebraical techniques of tensor theory.

3.2 Addition and scalar multiplication

If U and V are tensors of the same rank, with components $U_{\alpha\beta}$ and $V_{\alpha\beta}$ for example, and if (l_α) and (m_α) are two directions, then the expression

$$W_{\alpha\beta}l_\alpha m_\beta,$$

where

$$W_{\alpha\beta} = U_{\alpha\beta} + V_{\alpha\beta}$$

is clearly the sum of the two invariant bilinear functions,

$$U_{\alpha\beta}l_\alpha m_\beta \quad \text{and} \quad V_{\alpha\beta}l_\alpha m_\beta,$$

and is therefore a tensor, say W. We therefore *define* W to be the sum of the tensors U and V, and write

$$W = U + V = V + U.$$

Similarly if c is any number and

$$M_{\alpha\beta} = cU_{\alpha\beta},$$

then $M_{\alpha\beta}l_\alpha m_\beta = cU_{\alpha\beta}l_\alpha m_\beta$ is also an invariant bilinear function, and is therefore a tensor, say M. We *define* M to be the product of W by the number c, and write

$$M = cW.$$

Similar definitions manifestly apply to tensors of any rank.

3.3 Outer multiplication

If U and V are tensors, of ranks p and q respectively, then there is a tensor W, of rank $p + q$, which is called the 'outer product' of U and V. For example, if $p = 2$ and $q = 3$, and the components of U and V are $U_{\alpha\beta}$, $V_{\rho\sigma\tau}$, then we define the components of W by the formula

$$W_{\alpha\beta\rho\sigma\tau} = U_{\alpha\beta}V_{\rho\sigma\tau}.$$

To prove that these are the components of a tensor of rank 5, we note that

$$W_{\alpha\beta\rho\sigma\tau}\lambda_\alpha\mu_\beta l_\rho m_\sigma n_\tau = (U_{\alpha\beta}\lambda_\alpha\mu_\beta)(V_{\rho\sigma\tau}l_\rho m_\sigma n_\tau).$$

Hence the expression on the left hand side is an invariant multilinear function of the five directions (λ_α), (μ_α), (l_α), (m_α), (n_α), and is therefore a tensor.

Examples: If ρ is the density and (u_α) the velocity components of a fluid at a point P, then the numbers $\rho u_\alpha u_\beta$ are the components of a tensor of rank 2. This tensor describes the rate of convection of the (l_α)-component of momentum in the (m_α)-direction.

3.4 Spherical means of tensors and contraction

The physical significance of the process of 'contraction' is perhaps most easily explained by means of the concept of 'spherical means'.

In the case of a tensor

$$P = P(l_\alpha) = P_\alpha l_\alpha$$

of the first rank, the spherical mean is defined to be the average value of the function $P(l_\alpha)$ taken over the unit sphere

$$l_\alpha l_\alpha = 1.$$

Since to every direction (l_α) there corresponds an opposite direction $(-l_\alpha)$, the spherical mean $\bar{l}_\alpha$ of l_α, and therefore of P is clearly zero.

In the case of a tensor

$$P = P(l_\alpha, m_\alpha) = P_{\alpha\beta}l_\alpha m_\beta$$

of the second rank the spherical mean is defined to be the average value of the function $P(l_\alpha, l_\alpha)$ taken over the unit sphere

$$l_\alpha l_\alpha = 1.$$

Now the average value of $l_\alpha l_\beta$ is easily seen to be zero if $\alpha \neq \beta$, and to be equal to the average value of $\frac{1}{3}(l_1^2 + l_2^2 + l_3^2)$ if $\alpha = \beta$. Hence the spherical mean of P is

$$\tfrac{1}{3}(P_{11} + P_{22} + P_{33}) = \tfrac{1}{3}P_{\alpha\alpha}.$$

This average value is clearly an invariant scalar, i.e. it is a tensor of rank zero. In the case of the stress tensor $\sigma_{\alpha\beta}$ for a fluid, the spherical mean with the sign reversed

$$-(\sigma_{11} + \sigma_{22} + \sigma_{33})$$

is called the mean pressure at a point, even if the fluid is viscous.

$P_{\alpha\alpha}$ is the sum of the numbers in the principal diagonal of the matrix of P and is called the 'trace' of P (in German 'Spur' P), or by some writers, the 'divergence' of P.

In the general case of a tensor P of rank $p > 2$, we can form several spherical means. For example if P is a tensor of rank 3, and

$$P = P(l, m, n) = P_{\alpha\beta\gamma} l_\alpha m_\beta n_\gamma,$$

we can form the means

$$P_{\alpha\beta\beta} l_\alpha \overline{m_\beta m_\beta},\ P_{\alpha\beta\alpha}\, m_\beta \overline{l_\alpha l_\alpha},\ P_{\alpha\alpha\gamma} n_\gamma\, \overline{l_\alpha l_\alpha}.$$

Each of these is obviously a tensor of rank unity. The components of these tensors yield respectively,

$$\begin{array}{lcl} (P_{\alpha 11} + P_{\alpha 22} + P_{\alpha 33}), & & P_{\alpha\beta\beta} \\ (P_{1\alpha 1} + P_{2\alpha 2} + P_{3\alpha 3}), & \text{or} & P_{\beta\alpha\beta} \\ (P_{11\alpha} + P_{22\alpha} + P_{33\alpha}) & & P_{\beta\beta\alpha}. \end{array}$$

In the general case we have the theorem that if P is a tensor of rank $p > 2$ with components $P_{\alpha\beta\rho\sigma\tau\ldots}$ then the numbers $P_{\alpha\alpha\rho\sigma\tau\ldots}$ are the components of a tensor U of rank $p - 2$. The operation of equating two suffixes of a tensor and summing over all values of the repeated suffix is called 'contraction' (German 'Verjüngung').

An almost trivial but useful lemma (see § 4.4) is that if the components of the tensor R satisfy the equation

$$R_{\alpha\beta} u_\beta = 0,$$

for any vector u, then all the components of R are equal to zero.

Take $\alpha = 1$ and $u_\beta = (0, 1, 0)$, then

$$0 = R_{12},$$

and we can prove similarly that each component of R is zero.

Examples: (1) Verify directly that if $P_{\alpha\beta\gamma\delta}$ are the components of a fourth order tensor, then the set of numbers $P_{\alpha\alpha\gamma\delta}$ transforms like the components of a second order tensor.

(2) If $P_{\alpha\beta}$ and $Q_{\alpha\beta\sigma}$ are the components of two tensors of ranks 2 and 3 respectively, then the sets of numbers

$$R_{\beta\gamma\sigma} = P_{\alpha\beta}Q_{\alpha\gamma\sigma},$$
$$S_\sigma = P_{\alpha\beta}Q_{\alpha\beta\sigma}$$

are components of tensors of ranks 3 and 1 respectively. (These are examples of 'inner' multiplication, in which the operation of contraction is applied to an 'outer' product.)

(3) The 'Quotient Rule': if the set of numbers $P_{\alpha\beta}$ is such that, when $Q_\alpha l_\alpha$ is *any* tensor of rank unity, so also is

$$R_\alpha l_\alpha = Q_\alpha P_{\alpha\beta} l_\beta,$$

then the numbers $P_{\alpha\beta}$ are the components of a tensor of the second rank.

(4) Similarly, if the set of numbers $P_{\alpha\beta}$ is such that, when $Q = Q_{\alpha\beta} l_\alpha m_\beta$ is any tensor of rank two, so also is $R = P_{\alpha\beta} Q_{\beta\gamma} l_\alpha m_\gamma$, then the numbers $P_{\alpha\beta}$ are the components of a tensor of rank 2.

(5) Verify the transformation law for the set of numbers $P_{\alpha\beta}$ in examples 3 and 4.

3.5 Symmetry and antisymmetry

A tensor $P_{\alpha\beta\gamma\ldots}$ is said to be symmetrical or antisymmetrical in a pair of suffixes α and β according as

$$P_{\alpha\beta\gamma\ldots} = +P_{\beta\alpha\gamma\ldots}$$

or

$$P_{\alpha\beta\gamma\ldots} = -P_{\beta\alpha\gamma\ldots}$$

The unit tensor U with components $\delta_{\alpha\beta}$ is symmetrical in α and β. The alternating tensor A with components $\varepsilon_{\alpha\beta\gamma}$ is antisymmetrical in each pair of suffixes.

Now introduce the tensors S and A with components

$$S_{\alpha\beta\rho\sigma} = \tfrac{1}{2}(\delta_{\alpha\rho}\delta_{\beta\sigma} + \delta_{\alpha\sigma}\delta_{\beta\rho}),$$

$$A_{\alpha\beta\rho\sigma} = \tfrac{1}{2}\begin{vmatrix} \delta_{\alpha\rho} & \delta_{\alpha\sigma} \\ \delta_{\beta\rho} & \delta_{\beta\sigma} \end{vmatrix}. \qquad (3.5.1)$$

Then
$$S_{\alpha\beta\rho\sigma}P_{\rho\sigma} = \tfrac{1}{2}(P_{\alpha\beta} + P_{\beta\alpha})$$
and
$$A_{\alpha\beta\rho\sigma}P_{\rho\sigma} = \tfrac{1}{2}(P_{\alpha\beta} - P_{\beta\alpha}).$$

These equations show that the two sets of numbers

$$P_{(\alpha\beta)} = \tfrac{1}{2}(P_{\alpha\beta} + P_{\beta\alpha})$$
and
$$P_{[\alpha\beta]} = \tfrac{1}{2}(P_{\alpha\beta} - P_{\beta\alpha})$$

are each the components of a tensor. These tensors are called the symmetrical and antisymmetrical parts of $P_{\alpha\beta}$. Hence any tensor of the second rank can be expressed as the sum of a symmetric and an antisymmetric tensor.

Example: Construct tensors $S_{\alpha\beta\gamma\rho\sigma\tau}$ and $A_{\alpha\beta\gamma\rho\sigma\tau}$ such that

$$S_{\alpha\beta\gamma\rho\sigma\tau}P_{\rho\sigma\tau} = P_{\alpha\beta\gamma} + P_{\beta\gamma\alpha} + P_{\gamma\alpha\beta} + P_{\gamma\beta\alpha} + P_{\beta\alpha\gamma} + P_{\alpha\gamma\beta},$$
and
$$A_{\alpha\beta\gamma\rho\sigma\tau}P_{\rho\sigma\tau} = P_{\alpha\beta\gamma} + P_{\beta\gamma\alpha} + P_{\gamma\alpha\beta} - P_{\gamma\beta\alpha} - P_{\beta\alpha\gamma} - P_{\alpha\gamma\beta}.$$

3.6 Antisymmetric tensors of rank 2

In three dimensional space with any antisymmetric tensor P of rank 2 we can associate a tensor Q of rank 1 by means of the equations

$$\begin{aligned} Q_\alpha &= \tfrac{1}{2}\varepsilon_{\alpha\beta\gamma}P_{\beta\gamma}, \\ Q_1 &= P_{23} = -P_{32}, \\ Q_2 &= P_{31} = -P_{13}, \\ Q_3 &= P_{12} = -P_{21}. \end{aligned}$$

Now, by example 2 of § 2.2,

$$\begin{aligned} \tfrac{1}{2}\varepsilon_{\alpha\beta\gamma}P_{\beta\gamma}\varepsilon_{\rho\sigma\alpha} &= \tfrac{1}{2}(\delta_{\beta\rho}\delta_{\gamma\sigma} - \delta_{\beta\sigma}\delta_{\gamma\rho})P_{\beta\gamma} \\ &= \tfrac{1}{2}(P_{\rho\sigma} - P_{\sigma\rho}). \end{aligned}$$

Hence
$$P_{\rho\sigma} = \varepsilon_{\rho\sigma\alpha}Q_\alpha,$$
or
$$P_{\alpha\beta} = \begin{Vmatrix} 0 & Q_3 & -Q_2 \\ -Q_3 & 0 & Q_1 \\ Q_2 & -Q_1 & 0 \end{Vmatrix}$$

The determinant of the components of an antisymmetric tensor P of rank 2 is always zero.

3.7 Products of vectors

The outer product of two vectors u_α and v_α is the tensor $D_{\alpha\beta} = u_\alpha v_\beta$, commonly called a dyadic tensor or dyad.

The inner product, $u_\alpha v_\alpha$, is the 'scalar product', often written as $\mathbf{u}.\mathbf{v}$ or $(\mathbf{uv})$.

The antisymmetric part of the tensor $u_\alpha v_\beta$ yields the tensor $u_\alpha v_\beta - u_\beta v_\alpha$, with which is associated the vector

$$w_\alpha = \varepsilon_{\alpha\beta\gamma} u_\beta v_\gamma = \tfrac{1}{2}\varepsilon_{\alpha\beta\gamma}(u_\beta v_\gamma - u_\gamma v_\beta),$$

with components

$$w_1 = u_2 v_3 - u_3 v_2,$$
$$w_2 = u_3 v_1 - u_1 v_3,$$
$$w_3 = u_1 v_2 - u_2 v_1.$$

w_α is called the 'vector' product and is written as $\mathbf{u}\wedge\mathbf{v}$ or $[\mathbf{uv}]$.

3.8 The Chapman–Cowling notation

A very convenient notation for the application of tensor theory to the kinetic theory of gases has been developed by Chapman & Cowling.*

Vectors, as usual, are printed in Clarendon type, as **a**, **r**, etc., and the vector operator $(\partial/\partial x_a)$ is also treated as a vector and written as $\partial/\partial\mathbf{r}$.

Tensors of the second rank are printed in bold type, as **W**, **U**, etc. The transpose of the tensor **W** is written as $\overline{\mathbf{W}}$, and the symmetric part of the tensor **W** as $\overline{\overline{\mathbf{W}}}$, so that

$$\mathbf{W} = \tfrac{1}{2}(\overline{\mathbf{W}} + \overline{\overline{\mathbf{W}}}).$$

The trace of **W**, i.e. $W_{\alpha\alpha}$, is called its 'divergence'.

The 'deviation' of **W**, i.e. the non-divergent part of **W**, is defined as

$$\mathring{\mathbf{W}} = \mathbf{W} - \tfrac{1}{3}(W_{\alpha\alpha})\mathbf{U},$$

where **U** is the unit tensor with component $\delta_{\alpha\beta}$.

* Sydney Chapman & T. G. Cowling, *The Mathematical Theory of Non-Uniform Gases*, Cambridge University Press, 1st edition 1939, 2nd edition 1952.

Thus if **u** denotes the velocity of a fluid, we can form the velocity gradient tensor $\partial\mathbf{u}/\partial\mathbf{r}$, the rate of stress tensor $\overline{\overline{\partial\mathbf{u}/\partial\mathbf{r}}}$, and the rate of shear tensor $\overset{\circ}{\overline{\overline{\partial\mathbf{u}/\partial\mathbf{r}}}}$ (see Chapter IV).

The outer and inner products of two tensors **W** and **W**′ are called the 'single' and 'double' products and are written as

$$\mathbf{W}.\mathbf{W}' \quad \text{with components } w_\alpha w'_\beta,$$

and

$$\mathbf{W}:\mathbf{W}' \quad \text{with scalar given by } w_{\alpha\beta}w'_{\beta\alpha}.$$

Examples: (1) A rigid body is rotating about a fixed point P with angular velocity tensor ω. Show that its kinetic energy

$$\tfrac{1}{2}\Sigma\, m\dot{x}_\alpha\dot{x}_\alpha \quad \text{is} \quad \tfrac{1}{2}\omega_\alpha I_{\alpha\beta}\omega_\beta,$$

where I is the inertia tensor.

(2) If the angular momentum of the rigid body about P is defined to be the tensor H where

$$H_\alpha = \tfrac{1}{2}\Sigma\, m\varepsilon_{\alpha\beta\gamma}x_\beta\dot{x}_\gamma,$$

show that

$$H_\alpha = I_{\alpha\beta}\omega_\beta.$$

CHAPTER IV

The Calculus of Tensors

4.1 Introduction

One of the characteristic features of Cartesian tensors is the extreme simplicity of the rules for differentiation and their immediate applicability to the dynamics of continuous media, in particular to the construction of the strain tensor in elasticity and to the rate of strain tensor in fluid dynamics.

4.2 The differentiation of tensors

Hitherto we have mainly considered tensors which were in effect attached to a single point, as for example the inertia tensor for the centroid or any other prescribed point of a rigid body, or the angular velocity tensor referred to a prescribed origin. In the dynamics of continuous media, however, we have to consider tensors which are defined at every point of the region occupied by matter, and whose components are therefore functions of the space coordinates of the particle whose properties they describe. For example the stress tensor, $\sigma_{\alpha\beta}$, of a continuum, will vary from point to point and each component $\sigma_{\alpha\beta}$ is a function of the variables x_1, x_2, x_3.

In general a tensor P is a multilinear function of p direction cosines, (l_α), (m_α), . . . (where p is the rank of P), and

$$P = P_{\alpha\beta\ldots} l_\alpha m_\beta \ldots$$

The rate of change of P in the direction (λ_α) is clearly an invariant and is given by

$$U = \frac{\partial P}{\partial x_\rho} \lambda_\rho = \frac{\partial P_{\alpha\beta\ldots}}{\partial x_\rho} \lambda_\rho l_\alpha m_\beta \ldots$$

Hence the derivatives of $P_{\alpha\beta\ldots}$, the components of P, are themselves the components,

$$U_{\alpha\beta\ldots} = \partial P_{\alpha\beta\ldots}/\partial x_\rho \tag{4.2.1}$$

of a tensor of rank $(p + 1)$.

It is usual to write the derivatives of Cartesian tensors as

$$\partial P_{\alpha\beta\ldots}/\partial x_\rho = P_{\alpha\beta\ldots,\rho} \quad \text{or} \quad P_{\alpha\beta\ldots/\rho}$$

with a comma or solidus separating the suffix ρ which refers to the differentiation.

Example: Assuming the transformation law

$$Q_{ab} = R_{a\alpha}R_{b\beta}P_{\alpha\beta},$$

verify that $P_{\alpha\beta,\gamma}$ obeys the transformation law

$$Q_{ab,c} = R_{a\alpha}R_{b\beta}R_{c\gamma}P_{\alpha\beta,\gamma}.$$

4.3 Derived tensors

It is now an easy task to compile a list of the principal tensors formed by the processes of differentiation and contraction from any given tensor.

From any scalar ϕ we can form the gradient, grad ϕ, with components

$$\phi_{,\alpha} = \partial\phi/\partial x_\alpha,$$

and the Laplacian, which is the scalar

$$\Delta\phi = \phi_{,\alpha\alpha} = \phi_{,11} + \phi_{,22} + \phi_{,33}.$$

From any vector V_α we can form the second rank tensor,

$$V_{\alpha,\beta} = \partial V_\alpha/\partial x_\beta,$$

and thence, by contraction, the divergence of V_α,

$$\text{div } V = V_{\alpha,\alpha} = \frac{\partial V_1}{\partial x_1} + \frac{\partial V_2}{\partial x_2} + \frac{\partial V_3}{\partial x_3}.$$

We can also form the antisymmetric second rank tensor,

$$V_{\alpha,\beta} - V_{\beta,\alpha},$$

and the associated vector, the curl of V_α,

$$\begin{aligned}(\text{curl } V)_\alpha &= -\tfrac{1}{2}\varepsilon_{\alpha\beta\gamma}(V_{\beta,\gamma} - V_{\gamma,\beta}) \\ &= -\varepsilon_{\alpha\beta\gamma}V_{\beta,\gamma},\end{aligned}$$

with components

$$(\text{curl } V)_1 = \frac{\partial V_3}{\partial x_2} - \frac{\partial V_2}{\partial x_3},$$

$$(\text{curl } V)_2 = \frac{\partial V_1}{\partial x_3} - \frac{\partial V_3}{\partial x_1},$$

$$(\text{curl } V)_3 = \frac{\partial V_2}{\partial x_1} - \frac{\partial V_1}{\partial x_2}.$$

From any second rank tensor $P_{\alpha\beta}$ we can form a third rank tensor,

$$P_{\alpha\beta,\gamma} = \partial P_{\alpha\beta}/\partial x_\gamma,$$

and, by contraction, a *vector*, called the 'divergence' of P, viz.

$$(\text{div } P)_\alpha = P_{\alpha\beta,\beta} = \frac{\partial P_{\alpha 1}}{\partial x_1} + \frac{\partial P_{\alpha 2}}{\partial x_2} + \frac{\partial P_{\alpha 3}}{\partial x_3}.$$

Exercises: (1) Verify that curl grad $\phi = 0$, and that div curl $V = 0$.
(2) Show that curl curl $V =$ grad div $A - \nabla^2 V$.
(3) Show that div $(W \wedge V) = - W.\text{curl } V + V.\text{curl } W$.
(4) Show that curl $(\phi W) = \phi$ curl $W - U \wedge$ grad ϕ.

4.4 The strain tensor

An immediate application of tensor calculus is to the analytical description of the state of a solid body when strained by external forces. The most general description is obtained by specifying the displacement of each individual particle. We therefore suppose that the particle which was originally at the point P with coordinates x_α is displaced to the point $\bar{P}$ with coordinates $\bar{x}_\alpha$. A neighbouring point Q at $x_\alpha + \delta x_\alpha$ will be displaced to the point $\bar{Q}$, $\bar{x}_\alpha + \delta \bar{x}_\alpha$. There are various ways of defining the strain tensor, but the simplest is to form the invariant

$$I = \frac{|\overline{PQ}|^2 - |PQ|^2}{2|PQ|^2}, \qquad (4.4.1)$$

and to express it in terms of the direction cosines l_α of $\overrightarrow{PQ}$.

The displacement of the typical particle is given by the vector

$$u_\alpha = \bar{x}_\alpha - x_\alpha,$$

whence $$\frac{\partial \bar{x}_\alpha}{\partial x_\rho} = \delta_{\alpha\rho} + \frac{\partial u_\alpha}{\partial x_\rho} = \delta_{\alpha\rho} + u_{\alpha,\rho}.$$

Now $$\overline{PQ}^2 = \delta\bar{x}_\alpha\,\delta\bar{x}_\alpha = \bar{x}_{\alpha,\rho}\bar{x}_{\alpha,\sigma}\,\delta x_\rho\,\delta x_\sigma$$

$$PQ^2 = \delta s^2 = \delta x_\alpha\,\delta x_\alpha,$$

and $$l_\alpha = dx_\alpha/ds.$$

Hence $$I = \varepsilon_{\rho\sigma} l_\rho l_\sigma,$$

where $$\begin{aligned}\varepsilon_{\rho\sigma} &= \tfrac{1}{2}(\delta_{\alpha\rho} + u_{\alpha,\rho})(\delta_{\alpha\sigma} + u_{\alpha,\sigma}) - \tfrac{1}{2}\delta_{\rho\sigma} \\ &= \tfrac{1}{2}(u_{\sigma,\rho} + u_{\sigma,\rho}) + \tfrac{1}{2}(u_{\alpha,\rho}u_{\alpha,\sigma}).\end{aligned} \tag{4.4.2}$$

It is clear from the nature of the expression for $\varepsilon_{\rho\sigma}$ in terms of the derivatives of u_ρ that the numbers $\varepsilon_{\rho\sigma}$ form the components of a symmetric tensor of the second rank (and should not therefore be mistakable for the components of the alternating tensor $\varepsilon_{\rho\sigma\tau}$!). Typical components of $\varepsilon_{\rho\sigma}$ are the tensile strain parallel to the axis of x_1, viz.

$$\varepsilon_{11} = \frac{\partial u_1}{\partial x_1} + \frac{1}{2}\left\{\left(\frac{\partial u_1}{\partial x_1}\right)^2 + \left(\frac{\partial u_1}{\partial x_2}\right)^2 + \left(\frac{\partial u_1}{\partial x_3}\right)^2\right\},$$

and the shearing strain, in the plane of x_2, x_3, viz.

$$\varepsilon_{23} = \frac{1}{2}\left\{\frac{\partial u_2}{\partial x_3} + \frac{\partial u_3}{\partial x_2}\right\} + \frac{1}{2}\left\{\frac{\partial u_1}{\partial x_2}\frac{\partial u_1}{\partial x_3} + \frac{\partial u_2}{\partial x_2}\frac{\partial u_2}{\partial x_3} + \frac{\partial u_3}{\partial x_2}\frac{\partial u_3}{\partial x_3}\right\}.$$

The quadratic terms in curly brackets are often negligible compared with the linear terms, but they are of importance in the discussion of large deflexions of thin metal plates.

The linear terms,

$$\varepsilon_{\rho\sigma}{}^* = \tfrac{1}{2}(u_{\rho,\sigma} + u_{\sigma,\rho}), \tag{4.4.3}$$

are not independent of one another, but are connected by certain 'conditions of compatibility'. These are obtained by introducing the antisymmetric tensor

$$\omega_{\rho\sigma} = \tfrac{1}{2}(u_{\sigma,\rho} - u_{\rho,\sigma}). \tag{4.4.4}$$

We then find that

$$\begin{aligned}\omega_{\rho\sigma,\alpha} &= \tfrac{1}{2}(u_{\sigma,\rho\alpha} - u_{\rho,\sigma\alpha}) \\ &= \tfrac{1}{2}(u_{\sigma,\alpha\rho} + u_{\alpha,\sigma\rho}) \\ &\quad - \tfrac{1}{2}(u_{\rho,\alpha\sigma} + u_{\alpha,\rho\sigma}) \\ &= \varepsilon_{\sigma\alpha,\rho} - \varepsilon_{\rho\alpha,\sigma}. \end{aligned} \tag{4.4.5}$$

Hence

$$\omega_{\rho\sigma,\alpha\beta} = \varepsilon_{\sigma\alpha,\rho\beta} - \varepsilon_{\rho\alpha,\sigma\beta}$$

and

$$\omega_{\rho\sigma,\beta\alpha} = \varepsilon_{\sigma\beta,\rho\alpha} - \varepsilon_{\rho\beta,\sigma\alpha}. \tag{4.4.6}$$

Now the conditions of compatibility of equations (4.4.5) for the functions $\omega_{\rho\sigma}$ are simply that

$$\omega_{\rho\sigma,\alpha\beta} = \omega_{\rho\sigma,\beta\alpha}, \qquad (\alpha \neq \beta)$$

i.e., that

$$\varepsilon_{\sigma\alpha,\rho\beta} - \varepsilon_{\rho\alpha,\sigma\beta} = \varepsilon_{\sigma\beta,\rho\alpha} - \varepsilon_{\rho\beta,\sigma\alpha}. \tag{4.4.7}$$

These are the 'compatibility equations'. It is easily verified that there are six independent equations in (4.4.7) of which two typical equations are

$$\varepsilon_{22,33} + \varepsilon_{33,22} = 2\varepsilon_{23,23},$$

and

$$\varepsilon_{33,12} = \frac{\partial}{\partial x_3}(\varepsilon_{23,1} + \varepsilon_{31,1} - \varepsilon_{12,3}). \tag{4.4.8}$$

Exercises: (1) The strain carries the line element $\overrightarrow{PQ}$ into the line element $\overrightarrow{PQ}$. Show that the difference of these vectors has components $(\varepsilon_{\alpha\rho} - \omega_{\alpha\rho})\,\delta x_\rho + 0(\overline{|\delta x_\rho|}^2)$.

(2) Show that ε_{11} represents the proportional extension of a line element initially parallel to the PX_1 axis, and that $2\varepsilon_{23}$ represents the change in the angle between two line elements initially parallel to the axes of PX_2 and PX_3.

(3) The cubical dilatation is defined to be

$$\frac{\partial(u_1, u_2, u_3)}{\partial(x_1, x_2, x_3)} - 1 = \Delta = \varepsilon_{\alpha\alpha}.$$

If $\Delta = 0$, neglecting small quantities of the second order show that the basis at any point can be chosen so that

$$\varepsilon_{11} = 0,\ \varepsilon_{22} = 0,\ \varepsilon_{33} = 0.$$

4.5 The rate of strain tensor

A similar analysis applies to the motion of a fluid. Consider an element of the fluid near the point $P(x_\alpha)$. Let u_ρ be the velocity at any point. Then the velocity at a point $x_\alpha + h_\alpha$ is

$$u_\rho(x_\alpha + h_\alpha) = u_\rho(x_\alpha) + h_\alpha u_{\rho,\alpha}(x_\alpha) + 0(h_\alpha{}^2). \qquad (4.5.1)$$

Hence the relative velocity is

$$u_\rho(x_\alpha + h_\alpha) - u_\rho(x_\alpha) = h_\alpha \varepsilon_{\rho\alpha} - h_\alpha \omega_{\rho\alpha}.$$

where
$$\varepsilon_{\rho\alpha} = \tfrac{1}{2}(u_{\rho,\alpha\delta} + u_{\alpha,\rho})$$
and
$$\omega_{\rho\alpha} = \tfrac{1}{2}(u_{\alpha,\rho} - u_{\rho,\alpha}).$$

The tensor $\varepsilon_{\rho\alpha}$ represents the rate of strain at P, and the tensor $\omega_{\rho\alpha}$ represents the rate at which the fluid element is rotating about P. The vector associated with the antisymmetric tensor $2\omega_{\rho\alpha}$ is the 'vorticity' with components

$$\xi = \frac{\partial w}{\partial y} - \frac{\partial v}{\partial z}$$

$$\eta = \frac{\partial u}{\partial z} - \frac{\partial w}{\partial x} \qquad (u_\alpha = u, v, w)$$

$$\xi = \frac{\partial v}{\partial x} - \frac{\partial u}{\partial y}.$$

Examples: (1) Show that the total angular momentum in a small sphere of radius h about its centre P is

$$\frac{4\pi\rho}{15}(\text{curl } v)_P\, h^5 + 0(h^6),$$

where ρ is the fluid density, assumed to be uniform.

(2) Show that the proportional rate of increase in volume of any element is

$$\theta = \varepsilon_{\alpha\alpha}.$$

4.6 The momentum equations for a continuous medium

In constructing the equations of motion for a continuous medium we take the integral equations as fundamental and deduce the differential equations.

The integral equations refer to the material which is interior to a fixed closed surface S at a prescribed instant t, and they equate the rate of change of momentum to the resultant of the surface stresses and the body forces.

The total momentum is initially

$$\int \rho v_\alpha \, d\tau,$$

where ρ is the density, v_α the velocity and the integral is taken throughout the volume bounded by the surface S. Allowing for the flux of momentum across S, the rate of change in momentum is

$$\int \frac{\partial}{\partial t}(\rho v_\alpha) \, d\tau + \int \rho v_\alpha v_\beta l_\beta \, dS,$$

where (l_β) are the direction cosines of the outward drawn normal to S. By Green's transformation this reduces to the volume integral

$$\int \left\{ \frac{\partial}{\partial t}(\rho v_\alpha) + \frac{\partial}{\partial x_\beta}(\rho v_\alpha v_\beta) \right\} d\tau. \tag{4.6.1}$$

Similarly, if we equate the rate of decrease of the total mass of the material inside S to the total flux of material outwards across S, we obtain the integral form of the equation of continuity, viz.

$$-\int \frac{\partial \rho}{\partial t} \, d\tau = \int \rho v_\beta l_\beta \, dS,$$

whence

$$\int \left\{ \frac{\partial \rho}{\partial t} + \frac{\partial}{\partial x_\beta}(\rho v_\beta) \right\} d\tau = 0. \tag{4.6.2}$$

The stress $\sigma_{\alpha\beta}$ on an element of surface dS with its outward drawn normal in the direction l_β yields a force

$$\sigma_{\alpha\beta} l_\beta \, dS.$$

Hence the resultant of the surface stresses is the force

$$\int \sigma_{\alpha\beta} l_\beta \, dS = \int \frac{\partial \sigma_{\alpha\beta}}{\partial x_\beta} \, d\tau. \tag{4.6.3}$$

Hence, if the body forces have an intensity F_α per unit volume, the integral form of the momentum equation is, from (4.6.1) and (4.6.3),

$$\int\left\{\frac{\partial(\rho v_\alpha)}{\partial t}+\frac{\partial}{\partial x_\beta}(\rho v_\alpha v_\beta)-\frac{\partial\sigma_{\alpha\beta}}{\partial x_\beta}-F_\alpha\right\}d\tau=0. \qquad (4.6.4)$$

Since this is true for any volume we can deduce that the differential form of the momentum equation is

$$\frac{\partial(\rho v_\alpha)}{\partial t}+\frac{\partial(\rho v_\alpha v_\beta)}{\partial x_\beta}=\frac{\partial\sigma_{\alpha\beta}}{\partial x_\beta}+F_\alpha. \qquad (4.6.5)$$

The differential form of the equation of continuity (4.6.2) is

$$\frac{\partial\rho}{\partial t}+\frac{\partial(\rho v_\beta)}{\partial x_\beta}=0. \qquad (4.6.6)$$

By combining this equation with (4.6.5) we find that

$$\rho\left\{\frac{\partial v_\alpha}{\partial t}+v_\beta\frac{\partial v_\alpha}{\partial x_\beta}\right\}=\frac{\partial\sigma_{\alpha\beta}}{\partial x_\beta}+F_\alpha. \qquad (4.6.7)$$

The expression in curly brackets,

$$a_\alpha=\frac{\partial v_\alpha}{\partial t}+v_\beta\frac{\partial v_\alpha}{\partial x_\beta},$$

is clearly equal to

$$\frac{\partial v_\alpha}{\partial t}+\frac{\partial v_\alpha}{\partial x_\beta}\frac{dx_\beta}{dt}=\frac{dv_\alpha}{dt},$$

the absolute rate of change of v_α for a particle of fluid moving with velocity v_α. Hence a_α is the acceleration of a particle of the material. In the case of a solid not strained beyond the elastic limit the velocity v_α and its derivatives can be treated as small quantities whose squares and products can be neglected, so that the expression for the acceleration reduces to

$$a_\alpha=\partial v_\alpha/\partial t.$$

In the case of a fluid the full expression for the acceleration must be retained, except in the case of 'creeping flow' when v_α and $\partial v_\alpha/\partial x_\beta$ are everywhere small. The tensile stress $\sigma_{\alpha\beta}$ is usually replaced by the compressive stress

$$p_{\alpha\beta}=-\sigma_{\alpha\beta}$$

in fluid dynamics.

CHAPTER V

The Structure of Tensors

5.1 Introduction

For the theoretical physicist the most important tensors are undoubtedly the symmetric tensors of rank 2. All such tensors have the same structure, which is simple in character and of the greatest importance in linear algebra and in quantum theory. This structure is analysed in this chapter in terms of eigenvectors, eigenvalues and projection operators.

5.2 Projection operators

The very simplest type of symmetric tensor of rank 2 is the 'projection tensor' P, whose components $P_{\alpha\beta}$ are defined in terms of a direction λ_α by the equations

$$P_{\alpha\beta} = \lambda_\alpha \lambda_\beta. \tag{5.2.1}$$

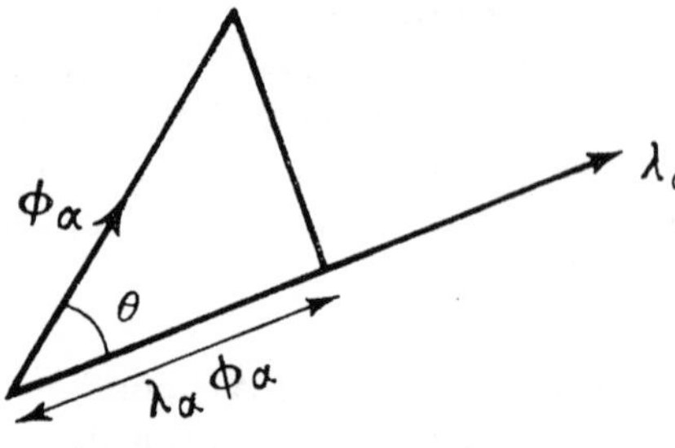

Consider the product of the tensor P and a vector ϕ. Since $P_{\alpha\beta}\phi_\beta = \lambda_\alpha(\lambda_\beta\phi_\beta)$, this product is a vector in the direction λ_α with a magnitude equal to the projection, $\lambda_\alpha\phi_\beta$, of the vector ϕ on the unit vector λ_β. This explains the name of the projection tensor, and implies its fundamental property that

$$P_{\alpha\beta}P_{\beta\gamma} = \lambda_\alpha\lambda_\beta\lambda_\beta\lambda_\gamma = \lambda_\alpha\lambda_\gamma \quad (\text{since } \lambda_\beta\lambda_\beta = 1)$$
$$= P_{\alpha\gamma}$$

or

$$P^2 = P,$$

for it is obvious that a repetition of the operation of projecting a vector on the same direction λ_α produces no further change.

On the other hand, if P and Q are projection tensors associated with two orthogonal directions (l_α) and (m_α), then

$$P_{\alpha\beta} = l_\alpha l_\beta, \quad Q_{\alpha\beta} = m_\alpha m_\beta,$$

and
$$P_{\alpha\beta} \cdot Q_{\beta\gamma} = l_\alpha l_\beta m_\beta m_\gamma = 0,$$

since
$$l_\beta m_\beta = 0.$$

Hence
$$PQ = 0,$$

the 'null' tensor, all of whose components are zero. Similarly $QP = 0$.

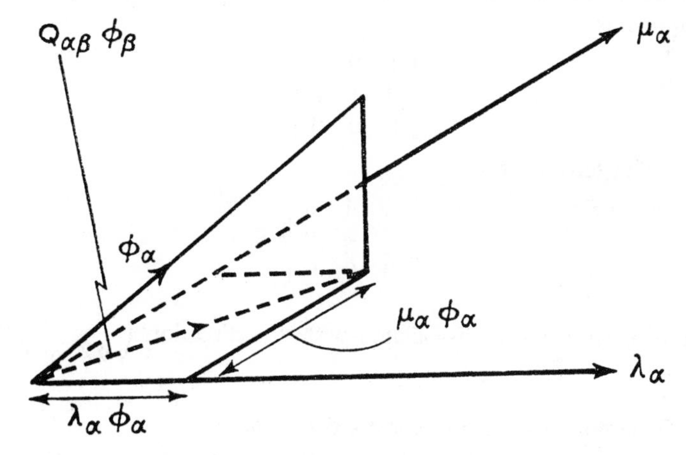

The tensor P gives the projection of a vector ϕ on the line with direction λ_α. We can also construct a tensor Q which gives the projection of a vector ϕ on the *plane* which contains the two orthogonal directions λ_α and μ_α. For if

$$Q_{\alpha\beta} = \lambda_\alpha\lambda_\beta + \mu_\alpha\mu_\beta,$$

then
$$Q_{\alpha\beta}\phi_\beta = \lambda_\alpha(\lambda_\beta\phi_\beta) + \mu_\alpha(\mu_\beta\phi_\beta)$$
= the α-component of the projection of ϕ on to the plane containing λ_α and μ_α.

Also $\qquad Q_{\alpha\beta}Q_{\beta\gamma} = \lambda_\alpha\lambda_\beta\lambda_\beta\lambda_\gamma + \lambda_\alpha\lambda_\beta\mu_\beta\mu_\gamma + \mu_\alpha\mu_\beta\lambda_\beta\lambda_\gamma + \mu_\alpha\mu_\beta\mu_\beta\mu_\gamma.$

But $\qquad \lambda_\beta\lambda_\beta = 1 = \mu_\beta\mu_\beta,$

and $\qquad \lambda_\beta\mu_\beta = 0.$

Hence $\qquad Q_{\alpha\beta}Q_{\beta\gamma} = Q_{\alpha\gamma}$

or $\qquad Q^2 = Q.$

Examples: (1) The projection tensor associated with the direction (0, 0, 1) has the components

$$\begin{Vmatrix} 0 & 0 & 0 \\ 0 & 0 & 0 \\ 0 & 0 & 1 \end{Vmatrix}$$

and the projection tensor associated with the plane $x_3 = 0$ has the components

$$\begin{Vmatrix} 1 & 0 & 0 \\ 0 & 1 & 0 \\ 0 & 0 & 0 \end{Vmatrix}$$

(2) The tensor with the matrix

$$\begin{Vmatrix} \frac{1}{2} & \frac{1}{2} & 0 \\ \frac{1}{2} & \frac{1}{2} & 0 \\ 0 & 0 & 0 \end{Vmatrix}$$

is a projection tensor associated with the direction $\left(\frac{1}{\sqrt{2}}, \frac{1}{\sqrt{2}}, 0\right)$

5.3 Definition of eigenvalues and eigenvectors

The central theorem which we shall obtain in this chapter is that any symmetric tensor S of rank 2 can be expressed in the canonical form

$$S = \lambda_1 P_{(1)} + \lambda_2 P_{(2)} + \lambda_3 P_{(3)}, \tag{5.3.1}$$

where λ_1, λ_2 and λ_3 are real numbers, and $P_{(1)}$, $P_{(2)}$, $P_{(3)}$ are projection tensors associated with three real orthogonal directions (l_α), (m_α), (n_α), the brackets being placed around the suffixes 1, 2, 3 to show that they do *not* indicate the components of a tensor! Any three vectors with the directions (l_α), (m_α), (n_α) respectively are called eigenvectors of S, and the numbers λ_1, λ_2, λ_3 are called the eigenvalues of S. The unit vectors $(\pm l_\alpha)$, $(\pm m_\alpha)$, $(\pm n_\alpha)$ are called normalized eigenvectors.

We note at once that theorem (5.3.1) implies that if we take the eigenvectors of S as a *basis*, then the projection tensors P_1, P_2, P_3 assume the simple forms

$$P_{(1)} = \begin{Vmatrix} 1 & 0 & 0 \\ 0 & 0 & 0 \\ 0 & 0 & 0 \end{Vmatrix} \quad P_{(2)} = \begin{Vmatrix} 0 & 0 & 0 \\ 0 & 1 & 0 \\ 0 & 0 & 0 \end{Vmatrix} \quad P_{(3)} = \begin{Vmatrix} 0 & 0 & 0 \\ 0 & 0 & 0 \\ 0 & 0 & 1 \end{Vmatrix}.$$

while S itself becomes

$$S = \begin{Vmatrix} \lambda_1 & 0 & 0 \\ 0 & \lambda_2 & 0 \\ 0 & 0 & \lambda_3 \end{Vmatrix}$$

These results show how the *structure* of the symmetric tensor S is revealed by introducing its eigenvectors and eigenvalues.

Assuming for the moment the truth of the theorem (5.3.1) we note that the eigenvector l_α must satisfy the equation

$$\begin{aligned} S_{\alpha\beta}l_\beta &= \lambda_1 P_{(1)\alpha\beta}l_\beta + \lambda_2 P_{(2)\alpha\beta}l_\beta + \lambda_3 P_{(3)\alpha\beta}l_\beta \\ &= \lambda_1 l_\alpha. \end{aligned}$$

Similarly $\quad S_{\alpha\beta}m_\beta = \lambda_2 m_\alpha,$

and $\quad S_{\alpha\beta}n_\beta = \lambda_3 n_\alpha. \qquad (5.3.2)$

Also
$$\begin{aligned} S^2 = SS &= (\lambda_1 P_{(1)} + \lambda_2 P_{(2)} + \lambda_3 P_{(3)})^2 \\ &= \lambda_1^2 P_{(1)} + \lambda_2^2 P_{(2)} + \lambda_3^2 P_{(3)}, \end{aligned}$$

since $\quad P_{(1)}P_{(3)} = P_{(3)}P_{(2)} = 0$, etc.

Similarly $\quad S^3 = \lambda_1^3 P_{(1)} + \lambda_2^3 P_{(2)} + \lambda_3^3 P_{(3)}$

and $\quad S^0 = P_{(1)} + P_{(2)} + P_{(3)}.$

To verify this last result we note that, if ϕ_α is any vector, then

$$\begin{aligned} (P_{(1)} + P_{(2)} + P_{(3)})\phi &= l_\alpha(l_\beta\phi_\beta) + m_\alpha(m_\beta\phi_\beta) + n_\alpha(n_\beta\phi_\beta) \\ &= \phi, \end{aligned}$$

since $(l_\beta\phi_\beta)$, $(m_\beta\phi_\beta)$, $(n_\beta\phi_\beta)$ are the projections of ϕ on to the three orthogonal directions given by l_α, m_α, n_α respectively. Hence

$$P_{(1)} + P_{(2)} + P_{(3)} = U,$$

where U is the unit tensor and, by convention,

$$S^\circ \equiv U.$$

Hence we can deduce that S satisfies the same algebraic equation as the eigenvalues λ_1, λ_2, λ_3. For these eigenvalues satisfy the equation

$$f(\lambda) \equiv (\lambda - \lambda_1)(\lambda - \lambda_2)(\lambda - \lambda_3) = 0.$$

Now $$f(S) \equiv (S - \lambda_1 U)(S - \lambda_2 U)(S - \lambda_3 U)$$
$$= S^3 - (\lambda_1 + \lambda_2 + \lambda_3)S^2 + (\lambda_2\lambda_3 + \lambda_3\lambda_1 + \lambda_1\lambda_2)S - \lambda_1\lambda_2\lambda_3 U$$
$$= \{\lambda_1^3 - (\lambda_1 + \lambda_2 + \lambda_3)\lambda_1^2 + (\lambda_2\lambda_3 + \lambda_3\lambda_1 + \lambda_1\lambda_2)\lambda_1 - \lambda_1\lambda_2\lambda_3\}P_{(1)} + \text{two similar terms in } P_{(2)}, P_{(3)}.$$

Therefore $$f(S) = f(\lambda_1)P_{(1)} + f(\lambda_2)P_{(2)} + f(\lambda_3)P_{(3)} = 0.$$

This is the Cayley–Hamilton equation.

Examples: (1) Verify that the eigenvalues and eigenvectors of the tensor

$$S = \begin{Vmatrix} 0 & C & 0 \\ C & 0 & C \\ 0 & C & 0 \end{Vmatrix}$$

where $$C = \tfrac{1}{2}\sqrt{2},$$

are $$\lambda_1 = -1, \lambda_2 = 0, \lambda_3 = +1,$$

and $$l_\alpha = (\tfrac{1}{2}, -C, \tfrac{1}{2}),$$
$$m_\alpha = (C, 0, -C),$$
$$n_\alpha = (\tfrac{1}{2}, C, \tfrac{1}{2}).$$

(2) Verify that the eigenvalues and eigenvectors of the tensor

$$S = \begin{Vmatrix} A + (C-A)l^2 & (C-A)lm & (C-A)ln \\ (C-A)ml & A + (C-A)m^2 & (C-A)mn \\ (C-A)nl & (C-A)nm & A + (C-A)n^2 \end{Vmatrix}$$

where A, B, C are constant and $l^2 + m^2 + n^2 = 1$, are

$$\lambda_1 = C, \lambda_2 = A, \lambda_3 = A,$$

and $l_\alpha = (l, m, n)$, with *any* pair of directions $(m_\alpha).(n_\alpha)$ orthogonal to one another and to l_α.

(3) In discussing the propagation of elastic waves in aeolotropic media, Lord Kelvin* showed that a very convenient solution of the equations

$$K_{\rho\sigma}\phi_\sigma = \lambda\phi_\rho \quad (\text{where } K_{\rho\sigma} = K_{\sigma\rho})$$

* Lord Kelvin, Baltimore Lectures, Cambridge University Press, 1904.

is obtained by writing

$$K_{23} = K_2K_3, \quad K_{31} = K_3K_1, \quad K_{12} = K_1K_2,$$
$$K_{11} = K_1^2 + L_{11}, \quad K_{22} = K_2^2 + L_{22}, \quad K_{33} = K_3^2 + L_{33},$$

or $$K_{\rho\sigma} = K_\rho K_\sigma + \delta_{\rho\sigma} L_{\rho\rho}.$$

Then the eigenvalues are the roots of the equation

$$\frac{K_1^2}{\lambda - L_{11}} + \frac{K_2^2}{\lambda - L_{22}} + \frac{K_3^2}{\lambda - L_{33}} = 1,$$

and the eigenvector corresponding to an eigenvalue λ is given by

$$\phi_\rho = \frac{K_\rho}{\lambda - L_{\rho\rho}} \quad \text{(not summed with respect to } \rho\text{).}$$

5.4 Existence of eigenvalues and eigenvectors

The existence of eigenvectors and eigenvalues can be discussed algebraically in terms of the system of simultaneous equations (5.3.2), but this discussion is complicated by the necessity of considering separately the cases in which the matrices of $S - \lambda_i U$ are of ranks 3, 2 or 1.* We therefore give an analytical discussion which is easily extended to tensors in n-dimensions, or to the theory of small oscillations of conservative dynamical systems.

The discussion turns on the values of the ratio

$$f(u) = \frac{S_{\alpha\beta}u_\alpha u_\beta}{u_\alpha u_\alpha} \tag{5.4.1}$$

for an arbitrary vector u.

We establish in succession the existence of eigenvalues

$$\lambda_1 \leqslant \lambda_2 \leqslant \lambda_3,$$

and their associated eigenvectors,

$$l_\alpha, \quad m_\alpha, \quad n_\alpha,$$

by determining the absolute minimum of $f(u)$ and certain conditional minima.

* If T is a matrix and r is the greatest integer such that at least one minor of order r is not zero, then T is said to be of 'rank' r.

If $\quad v_\alpha = | u_\alpha | (u_\beta u_\beta)^{-\frac{1}{2}}$ and $| S_{\alpha\beta} | \leqslant M$ for $\alpha, \beta = 1, 2, 3$,
then $| f(u) | \leqslant M(v_1 + v_2 + v_\alpha)^2$.

But $\quad 0 \leqslant v_\alpha \leqslant 1$, so that

$$-9M \leqslant f(u) \leqslant 9M,$$

and $f(u)$ is bounded. Also $f(u)$ is manifestly a continuous function of u_1, u_2, u_3. Hence $f(u)$ has a lower bound, say λ_1, which will be attained when

$$\partial f / \partial u_\alpha = 0,$$

i.e., when

$$S_{\alpha\beta} u_\beta = f(u) u_\alpha$$
$$= \lambda_1 u_\alpha. \qquad (5.4.2).$$

We have thus proved that the equations (5.4.2) must possess *a* solution when

$$\lambda_1 = \min f(u).$$

Let one unit vector satisfying these equations be

$$u_\alpha = l_\alpha.$$

To introduce the second eigenvector and eigenvalue we consider the values of $f(u)$ when u is restricted to be orthogonal to (l_α). $f(u)$ still remains a bounded and continuous function of u, and its conditional lower bound λ_2 will be attained when

$$\delta f = 0$$

subject to $u_\alpha l_\alpha = 0$,

i.e. when

$$S_{\alpha\beta} u_\beta \, \delta u_\alpha - f(u) u_\alpha \, \delta u_\alpha = 0$$

subject to

$$l_\alpha \, \delta u_\alpha = 0.$$

Hence the vector

$$S_\alpha \equiv S_{\alpha\beta} u_\beta - f(u) u_\alpha$$

is orthogonal to any vector (δu_α) which itself is orthogonal to (l_α). That is to say S_α is orthogonal to any pair of vectors which are themselves orthogonal to l_α. Hence S_α must be parallel to l_α, i.e. there exists a number c such that

$$S_\alpha = c l_\alpha$$

where c is a Lagrangian undetermined multiplier.

Hence

$$S_{\alpha\beta}u_\beta - f(u)u_\alpha = cl_\alpha,$$

i.e.

$$S_{\alpha\beta}u_\beta - \lambda_2 u_\alpha = cl_\alpha.$$

Therefore

$$l_\alpha S_{\alpha\beta}u_\beta - \lambda_2 l_\alpha u_\alpha = c(l_\alpha l_\alpha) = c.$$

But

$$l_\alpha u_\alpha = 0$$

and

$$l_\alpha S_{\alpha\beta}u_\beta = u_\beta S_{\beta\alpha}l_\alpha = \lambda_1 u_\beta l_\beta = 0.$$

Hence

$$c = 0,$$

and

$$S_{\alpha\beta}u_\beta = \lambda_2 u_\alpha. \tag{5.4.3}$$

We have thus proved that the equations (5.4.3) must possess a solution when

$$\lambda_2 = \min f(u),$$

subject to

$$u_\alpha l_\alpha = 0.$$

Let one unit vector satisfying these equations be

$$u_\alpha = m_\alpha.$$

Then

$$l_\alpha m_\alpha = 0.$$

Finally we introduce the direction n_α which is orthogonal to both l_α and m_α. Then

$$n_\alpha S_{\alpha\beta}l_\beta = \lambda_1 n_\alpha l_\alpha = 0,$$

and

$$n_\alpha S_{\alpha\beta}m_\beta = \lambda_2 n_\alpha m_\alpha = 0.$$

Hence the vector $n_\alpha S_{\alpha\beta}$ is orthogonal to both l_α and m_α. and is therefore parallel to n_α. Therefore there exists a number λ_3 such that

$$n_\alpha S_{\alpha\beta} = \lambda_3 n_\alpha,$$

and, since S is symmetric,

$$S_{\alpha\beta}n_\beta = \lambda_3 n_\alpha.$$

We have now established the existence of three orthogonal vectors l_α, m_α, n_α satisfying the equations

$$\begin{aligned} S_{\alpha\beta}l_\beta &= \lambda_1 l_\alpha, \\ S_{\alpha\beta}m_\beta &= \lambda_2 m_\alpha, \\ S_{\alpha\beta}n_\beta &= \lambda_3 n_\alpha. \end{aligned} \tag{5.4.4}$$

We can now prove that

$$S = \lambda_1 P_{(1)} + \lambda_2 P_{(2)} + \lambda_3 P_{(3)},$$

where $P_{(1)}$, $P_{(2)}$, $P_{(3)}$ are the projection operators associated with l_α, m_α, n_α.

Any vector u_α can be expressed in the form

$$u_\alpha = pl_\alpha + qm_\alpha + rn_\alpha,$$

where $$p = (u_\alpha l_\alpha), \quad q = (u_\alpha m_\alpha), \quad r = (u_\alpha n_\alpha),$$

as was proved in (1.4.2)

Now $$(\lambda_1 P_{(1)} + \lambda_2 P_{(2)} + \lambda_3 P_{(3)})_{\alpha\beta} u_\beta = \lambda_1 p l_\alpha + \lambda_2 q m_\alpha + \lambda_3 r n_\alpha$$
$$= pS_{\alpha\beta} l_\beta + qS_{\alpha\beta} m_\beta + rS_{\alpha\beta} n_\beta$$
$$= S_{\alpha\beta} u_\beta.$$

Hence $$S_{\alpha\beta} u_\beta = (\lambda_1 P_{(1)} + \lambda_2 P_{(2)} + \lambda_3 P_{(3)})_{\alpha\beta} u_\beta,$$

for *any* vector u_β.

Therefore by the lemma of § 3.4

$$S \equiv \lambda_1 P_{(1)} + \lambda_2 P_{(2)} + \lambda_3 P_{(3)}.$$

This is the canonical form of any symmetric tensor of rank 2. The numbers λ_1, λ_2, λ_3 are called its eigenvalues, and the directions l_α, m_α, n_α are called its eigenvectors.

5.5 The secular equation

We have shown in the preceding section that any eigenvalue λ of S and its associated eigenvector k_α satisfy the equation

$$S_{\alpha\beta} k_\beta = \lambda k_\alpha = \lambda \delta_{\alpha\beta} k_\beta.$$

These simultaneous linear equations for k_1, k_2, k_3 are therefore consistent, and hence the determinant $\Delta(\lambda)$ of

$$S - \lambda U$$

must vanish. The equation

$$\Delta(\lambda) = 0 \quad \text{or} \quad \begin{vmatrix} S_{11}-\lambda & S_{12} & S_{13} \\ S_{21} & S_{22}-\lambda & S_{23} \\ S_{31} & S_{32} & S_{33}-\lambda \end{vmatrix} = 0$$

has therefore three roots $\lambda = \lambda_1, \lambda_2, \lambda_3$, and must be equivalent to the equation

$$(\lambda_1 - \lambda)(\lambda_2 - \lambda)(\lambda_3 - \lambda) = 0.$$

Hence
$$\begin{aligned} \lambda_1 + \lambda_2 + \lambda_3 &= S_{11} + S_{22} + S_{33} \\ &= S_{\alpha\alpha}, \\ \lambda_2\lambda_3 + \lambda_3\lambda_1 + \lambda_1\lambda_2 &= (S_{22}S_{33} - S_{23}{}^2) + \ldots \\ &= \tfrac{1}{2}(S_{\alpha\alpha})^2 - \tfrac{1}{2}(S_{\alpha\beta}S_{\beta\alpha}), \end{aligned}$$
and
$$\lambda_1\lambda_2\lambda_3 = \det S.$$

Examples: (1) If $C_{\alpha\beta}$ denote the cofactor of $S_{\alpha\beta} - \lambda\delta_{\alpha\beta}$ in the determinant $\Delta(\lambda)$, then the numbers $C_{\alpha\beta}$ are the components of a symmetric tensor of rank 2.

(2)
$$\begin{aligned} C_{\alpha\beta}l_\beta &= (\lambda_2 - \lambda)(\lambda_3 - \lambda)l_\alpha, \\ C_{\alpha\beta}m_\beta &= (\lambda_3 - \lambda)(\lambda_1 - \lambda)m_\alpha, \\ C_{\alpha\beta}n_\beta &= (\lambda_1 - \lambda)(\lambda_2 - \lambda)n_\alpha. \end{aligned}$$

(3) If the determinant $\Delta(\lambda)$ has a repeated zero λ, then λ is also a zero of each of the minors of $\Delta(\lambda)$.

(4) As an example of a non-symmetric tensor consider the transformation tensor R which carries the base (PX_α) into the base specified by the three orthogonal unit vectors $(a_\alpha)(b_\alpha)(c_\alpha)$. Then

$$R = \begin{Vmatrix} a_1 & a_2 & a_3 \\ b_1 & b_2 & b_3 \\ c_1 & c_2 & c_3 \end{Vmatrix}$$

The characteristic equation is

$$0 = \det(R - cU) = -c^3 + 1 + (c^2 - c)(a_1 + b_2 + c_3).$$

Now by example 3, § 1.9,

$$a_1 + b_2 + c_3 = 1 + 2\cos\omega,$$

where ω is the angle through which R rotates any vector about the axis of R. Hence the roots of the characteristic equation are

$$c = 1, \quad e^{i\omega}, \quad e^{-i\omega}.$$

The equation $R_{\alpha\beta}u_\beta = u_\alpha$ is also satisfied, of course, by the vector $(\lambda_1, \lambda_2, \lambda_3)$ along the axis of R.

(5) The eigenvalues and eigenvectors of the antisymmetric tensor

$$A = \left\| \begin{matrix} 0 & c & -b \\ -c & 0 & a \\ b & -a & 0 \end{matrix} \right\|$$

are $\quad = 0, \quad \pm\, i(a^2 + b^2 + c^2),$ and $u_\alpha, p_\alpha + iq_\alpha,$
where $u_\alpha = (a, b, c),$
and p_α, q_α are any pair of vectors orthogonal to u_α and to one another.

(6) Show that the eigenvalues and eigenvectors of Maxwell's stress tensor $F_{\alpha\beta}$ for an electrostatic field are

$$\lambda = (8\pi)^{-1}E_\sigma E_\sigma, \quad \lambda = -\,(8\pi)^{-1}E_\sigma E_\sigma \text{ (twice)}, \quad \text{and} \quad u_\alpha = E_\alpha,$$

together with any pair of vectors orthogonal to E_α.

$$[4\pi F_{\alpha\beta} = E_\alpha E_\beta - \tfrac{1}{2}\delta_{\alpha\beta}(E_\sigma E_\sigma).]$$

CHAPTER VI

Isotropic Tensors

6.1 Introduction

It is, of course, impossible to excogitate the fundamental laws of a physical system from a purely mathematical investigation such as the theory of tensors. Nevertheless tensor theory serves two useful purposes for the physicist. We have already seen that many fundamental characteristics of a physical system, such as the moments and products of inertia of a rigid body, or the stresses and strains in an elastic body, are appropriately *represented* by tensors. In this chapter we shall show that tensor theory can also determine with precision the simplest possible relations between different tensors, and, in particular can determine the simplest constitutive relations of elastic bodies and viscous fluids.

Such hypothetical relations must of course be tested by experiment, but it is of the greatest advantage to the experimenter to know what are the simplest possible forms of the physical laws which he proposes to discover. Not that there is any reason to suppose that Nature prefers simplicity, but that simpler hypotheses are more easily tested than complex theories.

6.2 Definition of isotropic tensors

It will readily be admitted that the simplest possible relations between tensors must be *linear*. For example the simplest possible relation between the stress tensor $\sigma_{\alpha\beta}$ and the strain tensor $\varepsilon_{\alpha\beta}$ must be of the form

$$\sigma_{\alpha\beta} = c_{\alpha\beta\rho\sigma}\varepsilon_{\rho\sigma}. \tag{6.2.1}$$

Now if the elastic material is homogeneous and isotropic, its properties will be the same at all points and in all forms of reference. Hence the equations (6.2.1) must remain *invariant* under a transformation of base R, i.e. under a rotation of the frame of reference.

Now under such a transformation the tensors σ and ε become $\bar{\sigma}$ and $\bar{\varepsilon}$ where

$$\bar{\sigma}_{ab} = R_{a\alpha}\sigma_{\alpha\beta}R'_{\beta b} \quad \text{by (2.5.2)}$$

and
$$\bar{\varepsilon}_{rs} = R_{r\rho}\varepsilon_{\rho\sigma}R'_{\sigma s}$$

or
$$\varepsilon_{\rho\sigma} = R'_{\rho r}\bar{\varepsilon}_{rs}R_{s\sigma}$$

Hence
$$\begin{aligned}\bar{\sigma}_{ab} &= R_{a\alpha}c_{\alpha\beta\rho\sigma}R'_{\rho r}\bar{\varepsilon}_{rs}R_{s\sigma}R'_{\beta b}\\ &= \bar{c}_{abrs}\bar{\varepsilon}_{rs},\end{aligned}$$

where
$$\bar{c}_{abrs} = R_{a\alpha}R_{b\beta}R_{r\rho}R_{s\sigma}c_{\alpha\beta\rho\sigma}.$$

The coefficients $\bar{c}_{\alpha\beta\rho\sigma}$ are therefore the components of a tensor of rank 4. But if the stress-strain equations remain invariant, the coefficients must remain unaltered, i.e.

$$\bar{c}_{abrs} = c_{abrs}.$$

Thus the tensor $\bar{c}_{abrs}$ must have the *same set* of coefficients in all bases. Such tensors are called 'isotropic' tensors, and they will clearly play a fundamental role in the formulation of the simplest physical relations for isotropic substances.

In general we define an isotropic tensor of any rank by the criterion that its components form the same set of numbers in *all* bases, or that it is invariant under *any* rotation.

We have already encountered a number of isotropic tensors:

$$\delta_{\alpha\beta}, \quad \varepsilon_{\alpha\beta\gamma}, \quad (\S\ 2.2)$$

$$\left.\begin{aligned}S_{\alpha\beta\rho\sigma} &= \tfrac{1}{2}(\delta_{\alpha\rho}\delta_{\beta\sigma} + \delta_{\alpha\sigma}\delta_{\beta\rho})\\ A_{\alpha\beta\rho\sigma} &= \tfrac{1}{2}(\delta_{\alpha\rho}\delta_{\beta\sigma} - \delta_{\alpha\sigma}\delta_{\beta\rho})\end{aligned}\right\} (\S 3.5.1)$$

and we now proceed to determine *all* the isotropic tensors of ranks 1, 2, 3 and 4.

Now if $T_{\alpha\beta\gamma\ldots}$ are the components of an isotropic tensor of rank p then the multilinear form

$$T = T_{\alpha\beta\gamma\ldots}l_\alpha m_\beta n_\gamma \ldots$$

in p vectors l_α, m_β, n_γ . . . remains *invariant* when these vectors are subjected to any rotation R. Hence the enumeration of the isotropic tensors is equivalent to the enumeration of the multilinear rotational invariants.

The investigation in this chapter is restricted to isotropic tensors of rank not exceeding four. Even so it threatens to become tedious although it is systematic and strictly elementary. By way of excuse and introduction we draw attention to the following difficulty:

When we consider the invariants of three tensors, say x_α, y_α, z_α, we can easily prove by an appeal to elementary geometry that *any* invariant is necessarily a function of the six quadratic invariants.

$$(x_\alpha x_\alpha),\quad (y_\alpha y_\alpha),\quad (z_\alpha z_\alpha),\quad (y_\alpha z_\alpha),\quad (z_\alpha x_\alpha),\quad (x_\alpha y_\alpha).$$

This however is insufficient to determine the *multilinear* invariants, which are not all *linear* functions of the quadratic invariants, as is evidenced by the simple example

$$I = \varepsilon_{\alpha\beta\gamma} x_\alpha y_\beta z_\gamma = \begin{vmatrix} x_1 & x_2 & x_3 \\ y_1 & y_2 & y_3 \\ z_1 & z_2 & z_3 \end{vmatrix}$$

where

$$I^2 = \begin{vmatrix} (x_\alpha x_\alpha) & (x_\alpha y_\alpha) & (x_\alpha z_\alpha) \\ (y_\alpha x_\alpha) & (y_\alpha y_\alpha) & (y_\alpha z_\alpha) \\ (z_\alpha x_\alpha) & (z_\alpha y_\alpha) & (z_\alpha z_\alpha) \end{vmatrix}$$

It is therefore necessary to emphasize that the multilinear invariants which we seek are *polynomials* in the vector components. The following exposition is based upon the far more general theory developed by Weyl.

6.3 Isotropic tensors in two dimensions

Weyl's theory provides an inductive process by which we can successively enumerate the polynomial invariants in 1, 2, 3, . . . dimensions, and prove that *all* the invariants which are polynomials in the components of p vectors, x, y, u, v, . . . , in any number of dimensions are polynomials in the $\frac{1}{2}p(p+1)$ scalar products (xx), (xy) . . . and in the 'bracket' products $[xyu \ldots]$. These latter are the determinants of n rows and n columns, the rows consisting of the components (in order) of n different vectors chosen from x, y, u, v, . . .

We, however, shall make a flying start by starting with invariants in two dimensions. Under a typical rotation R a vector x with components (x_1, x_2) is transformed into a vector with components $\bar{x}_1, \bar{x}_2$ where

$$\begin{aligned} \bar{x}_1 &= x_1 \cos\theta + x_2 \sin\theta, \\ \bar{x}_2 &= -x_1 \sin\theta + x_2 \cos\theta. \end{aligned} \tag{6.3.1}$$

It is however much more convenient to work with the complex components

$$\begin{aligned} x_+ &= x_1 + ix_2, \quad x_- = x_1 - ix_2 \\ \bar{x}_+ &= \bar{x}_1 + i\bar{x}_2, \quad \bar{x}_- = \bar{x}_1 - i\bar{x}_2. \end{aligned}$$

The transformation law (6.3.1) then becomes

$$\begin{aligned} \bar{x}_+ &= \tau x_+, \\ \bar{x}_- &= \tau^{-1} x_- \end{aligned}$$

where $$\tau = \cos\theta - i\sin\theta.$$

The polynomial invariants of a single vector x must be polynomials in x_+, x_-, and it is manifest that the criterion of invariance compels these latter to be polynomials in the product

$$x_+x_- = x_1^2 + x_2^2.$$

Hence there are no linear invariants, nor indeed any invariants of odd degree in the components of a single vector.

When we proceed to consider a second vector y with complex components y_+, y_- our polynomial invariants will contain terms of the type

$$x_+^\alpha x_-^\beta y_+^\rho y_-^\sigma,$$

and such a term will be transformed into

$$\bar{x}_+^\alpha \bar{x}_-^\beta \bar{y}_+^\rho \bar{y}_-^\sigma = \tau^{\alpha-\beta+\rho-\sigma} x_+^\alpha x_-^\beta y_+^\rho y_-^\sigma.$$

Hence such a term cannot appear in an invariant polynomial unless

$$\alpha - \beta + \rho - \sigma = 0. \tag{6.3.2}$$

Therefore the degree s of such a tensor is

$$s = \alpha + \beta + \rho + \sigma = 2(\alpha + \rho),$$

and is therefore necessarily *even*.

Now if $\alpha \geqslant \beta$, then by (6.3.2), $\rho \leqslant \sigma$, and we can write the typical term

$$x_+^\alpha x_-^\beta y_+^\rho y_-^\sigma$$

in the form $$(x_+x_-)^\beta (y_+y_-)^\rho (x_+y_-)^{\alpha-\beta}.$$

Alternatively if $\alpha \leqslant \beta$, then $\rho \geqslant \sigma$, and the corresponding form is

$$(x_+x_-)^\alpha(y_+y_-)^\sigma(x_-y_+)^{\beta-\alpha}.$$

But $$x_+y_- = (xy) - i[xy]$$
and $$x_-y_+ = (xy) + i[xy],$$
where $$(xy) = x_1y_1 + x_2y_2,$$
and $$[xy] = x_1y_2 - x_2y_1.$$

We therefore conclude that any invariant polynomial in the components of two vectors x, y is a polynomial in the scalar products

$$(xx), \quad (yy), \quad (xy),$$

and in the vector or 'bracket' product $[xy]$. Moreover, since

$$[xy]^2 = (xx)(yy) - (xy)^2,$$

any even power of $[xy]$ can be expressed in terms of (xx), (yy) and (xy), while any odd power of $[xy]$ can be expressed as the product of $[xy]$ and powers of (xx), (yy) and (xy). Therefore any invariant polynomial can be expressed as a polynomial in (xx), (yy), (xy) and $[xy]$, which is of the first degree in $[xy]$.

The only *multilinear* invariants are obviously

$$(xy) \quad \text{and} \quad [xy],$$

and the corresponding isotropic tensors in 2 dimensions are numerical multiples of

$$\begin{vmatrix} 1 & 0 \\ 0 & 1 \end{vmatrix} \quad \text{and} \quad \begin{vmatrix} 0 & 1 \\ -1 & 0 \end{vmatrix}$$

6.4 Isotropic tensors of rank 2 in three dimensions

We can now take the next step in the inductive process. It is convenient to use the notation $A(m, n)$, $B(m, n)$, . . . $I(m, n)$ to denote invariants in the components of m vectors in n dimensions, which are linear in the components of each vector. Thus $A(m, 2)$ will involve only the components x_1, x_2, y_1, y_2 . . . of the m vectors x, y, . . . while $A(m, 3)$ will also involve the components x_3, y_3, . . . We shall also distinguish the multilinear invariants in the first and second components of the vectors by asterisks, so that

$$(xy)^* = x_1y_1 + x_2y_2,$$
$$[xy]^* = x_1y_2 - x_2y_1,$$

and

$$(xy) = (xy)^* + x_3y_3.$$

Then any invariant which is bilinear in three dimensions can be expressed in the form

$$I(2, 3) = I(2, 2) + x_3A(1, 2) + y_3B(1, 2) + x_3y_3C(0, 2).$$

But there are no linear invariants in two dimensions, so that

$$A(1, 2) = 0, \quad B(1, 2) = 0.$$

The invariant $C(0, 2)$ of degree zero is a mere constant C, and $I(2, 2)$ is a linear function of $(xy)^*$ and $[xy]^*$ by the result of § 6.3. Therefore

$$I(2, 3) = a(xy)^* + b[xy]^* + cx_3y_3,$$

where a, b, c are constants.

It is now obvious that $I(2, 3)$ will be invariant for rotation in three dimensions only if $a = c$ and $b = 0$. Hence

$$I(2, 3) = c(xy) = c\delta_{\alpha\beta}x_\alpha y_\beta,$$

and the only isotropic tensors of rank 2 in three dimensions are numerical multiples of $\delta_{\alpha\beta}$.

6.5 Isotropic tensors of rank 3 in three dimensions

An invariant which is multilinear in the components of three vectors x, y, z can be expressed in the form

$$I(3; 3) = J(x, y, z; 2) + x_3A(y, z; 2) + y_3B(z, x; 2) + z_3C(x, y; 2) + y_3z_3P(x; 1) + z_3x_3Q(y; 1) + x_3y_3R(z; 1) + x_3y_3z_3S,$$

where we have noted explicitly the vectors which figure in the invariants $A(y, z; 2), \ldots, P(x; 1), \ldots$ together with the number of dimensions.

Now $J(x, y, z; 2)$ is an invariant in the components $x_1, x_2, y_1, y_2, z_1, z_2$, and by the arguments employed in § 6.3 it follows that J is a polynomial in $(yz)^*$, $(zx)^*$, $(xy)^*$, $[yz]^*$, $[zx]^*$ and $[xy]^*$. Hence the degree of J must be even. But J is to be linear in the components of each vector and its degree must therefore be odd. Therefore J must be zero. A similar argument applies to P, Q and R.

We can also infer that $A(y, z; 2)$ must be linear in $(yz)^*$ and $[yz]^*$, so that $I(3, 3)$ must be linear in the following seven arguments:

$$\begin{gathered} x_3(yz) - x_3y_3z_3, \quad x_3[yz]^*, \\ y_3(zx) - y_3z_3x_3, \quad y_3[zx]^*, \\ z_3(xy) - z_3x_3y_3, \quad z_3[xy]^*, \\ \text{and } x_3y_3z_3. \end{gathered}$$

Finally we apply the criterion that $I(3, 3)$ is invariant for all rotations in three dimensions; indeed, it is sufficient to require that $I(3, 3)$ is invariant under the rotations

$$\begin{aligned} &(1)\ x_1 \to x_1,\ x_2 \to x_3,\ x_3 \to -x_1; \\ &(2)\ x_1 \to -x_3,\ x_2 \to x_2,\ x_3 \to x_1; \\ &(3)\ x_1 \to x_2;\ x_2 \to -x_1;\ x_3 \to x_3; \\ &(4)\ x_1 \to x_2;\ x_2 \to x_3;\ x_3 \to x_1. \end{aligned}$$

We thus find after a simple calculation that $I(3, 3)$ must be a numerical multiple of

$$x_3[yz]^* + y_3[zx]^* + z_3[xy]^* = \begin{vmatrix} x_1 & x_2 & x_3 \\ y_1 & y_2 & y_3 \\ z_1 & z_2 & z_3 \end{vmatrix} = \varepsilon_{\alpha\beta\gamma}x_\alpha y_\beta z_\gamma.$$

Thus the only isotropic tensors of rank 3 in three dimensions are numerical multiples of the alternating tensor $\varepsilon_{\alpha\beta\gamma}$.

6.6 Isotropic tensors of rank 4 in three dimensions

An invariant which is multilinear in the components of four vectors x, y, z, u can be expressed in the form

$$\begin{aligned} I(4; 4) = {} & J(x, y, u, v; 2) \\ & + x_3A(y, u, v; 2) + \dots \\ & + y_3z_3L(u, v; 2) + \dots \\ & + y_3z_3u_3S(v; 2) + \dots \\ & + x_3y_3z_3u_3K, \end{aligned}$$

where A, L, S, etc., represents invariants in two dimensions in the vectors indicated.

As before the invariants A, S, etc., must all be zero; $L(u, v; 2)$ must be linear in (uv) and in $[uv]^*$; while $J(x, y, u, v; 2)$ must be linear in $(xy)^*(uv)^*$,$(xu)^*$, $(yv)^*$, $(xv)^*(yu)^*$. Now

$$(xy)^*(uv)^* = (xy)(uv) - x_3y_3(uv) - u_3v_3(xy) - x_3y_3u_3v_3, \text{ etc.}$$

We are therefore left with a linear function of

(1) $(xy)(uv)$ and 2 similar terms,
(2) $(xy)u_3v_3$ and 5 similar terms,
(3) $x_3y_3u_3v_3$,
(4) $x_3y_3(uv)$ and 5 similar terms, and
(5) $x_3y_3\,uv$ and 5 similar terms.

Lastly, as in § 6.5, we apply the criterion that $I(4; 4)$ is invariant for all rotations. This enables us to eliminate all the arguments listed under (2), (3), (4) and (5). We can thus conclude that $I(4; 4)$ is a linear function of

$$(xy)(uv), \quad (xu)(yv) \quad \text{and} \quad (xv)(yu),$$

i.e. of $\delta_{\alpha\beta}\delta_{\rho\sigma}x_\alpha y_\beta u_\rho v_\sigma$, of $\delta_{\alpha\rho}\delta_{\beta\sigma}x_\alpha y_\beta u_\rho v_\sigma$

and of $\delta_{\alpha\sigma}\delta_{\beta\rho}x_\alpha y_\beta u_\rho v_\sigma$.

Hence the only isotropic tensors of rank 4 in three dimensions are linear functions of

$$\delta_{\alpha\beta}\delta_{\rho\sigma}, \quad \delta_{\alpha\rho}\delta_{\beta\sigma}, \quad \delta_{\alpha\sigma}\delta_{\beta\rho}.$$

6.7 The stress–strain relations for an isotropic elastic medium

We can now solve the problem propounded in § 6.2 of constructing explicitly the simplest relation between the stress tensor $\sigma_{\alpha\beta}$ and the strain tensor $\varepsilon_{\alpha\beta}$ – or, if we wish to be slightly more general, between the stress tensor $\sigma_{\alpha\beta}$ and the tensor $u_{\alpha,\beta}$. This relation must be of the form

$$\sigma_{\alpha\beta} = c_{\alpha\beta\rho\sigma}u_{\rho,\sigma}, \tag{6.7.1}$$

where $c_{\alpha\beta\rho\sigma}$ is an isotropic tensor of rank 4. Now we have just shown in § 6.6 that $c_{\alpha\beta\rho\sigma}$ must have the form

$$c_{\alpha\beta\rho\sigma} = A\delta_{\alpha\beta}\delta_{\rho\sigma} + B\delta_{\alpha\rho}\delta_{\beta\sigma} + C\delta_{\alpha\sigma}\delta_{\beta\rho}.$$

Hence the stress–strain relation must be

$$\sigma_{\alpha\beta} = A\delta_{\alpha\beta}u_{\rho,\rho} + Bu_{\alpha,\beta} + Cu_{\beta,\alpha}.$$

But $\sigma_{\alpha\beta}$ is symmetrical in α and β. Hence

$$B = C.$$

Now $$\varepsilon_{\alpha\beta} = \tfrac{1}{2}(u_{\alpha,\beta} + u_{\beta,\alpha})$$

and the dilatation $$\Delta = \varepsilon_{\alpha,\alpha} = u_{\alpha,\alpha}.$$

Therefore $$\sigma_{\alpha\beta} = A\delta_{\alpha\beta}\Delta + 2B\varepsilon_{\alpha\beta}.$$

In the usual notation $A \equiv \lambda$ and $B = \mu$, so that

$$\begin{aligned} \sigma_{11} &= \lambda\Delta + 2\mu\varepsilon_{11}, & \sigma_{23} &= 2\mu\varepsilon_{23} \\ \sigma_{22} &= \lambda\Delta + 2\mu\varepsilon_{22}, & \sigma_{31} &= 2\mu\varepsilon_{31} \\ \sigma_{33} &= \lambda\Delta + 2\mu\varepsilon_{33}, & \sigma_{12} &= 2\mu\varepsilon_{12}. \end{aligned} \tag{6.7.2}$$

A comparison with standard treatises on elasticity, with proper allowance for different notations, shows that μ is the coefficient of rigidity. The other coefficient λ has no special name, but enters into the formulae for

(1) the modulus of compression $k = \lambda + \frac{2}{3}\mu$,

(2) Young's modulus $E = \dfrac{\mu(3\lambda + 2\mu)}{\lambda + \mu}$, and

(3) Poisson's ratio $\sigma = \dfrac{\lambda}{2(\lambda + \mu)}$.

The inverse relations are easily found from (6.7.2) in the form

$$\begin{aligned} \varepsilon_{11} &= E^{-1}\{\sigma_{11} - \sigma({}_{22} + \sigma_{33})\}, \\ \varepsilon_{22} &= E^{-1}\{\sigma_{22} - \sigma(\sigma_{33} + \sigma_{11})\}, \\ \varepsilon_{33} &= E^{-1}\{\sigma_{33} - \sigma(\sigma_{11} + \sigma_{22})\}. \end{aligned} \tag{6.7.3}$$

Example: Show that there exists a strain energy function

$$W = \tfrac{1}{2}\lambda\Delta^2 + \mu(\varepsilon_{\alpha\beta}\varepsilon_{\alpha\beta}),$$

such that $$\partial W/\partial\varepsilon_{\alpha\beta} = \sigma_{\alpha\beta}.$$

6.8 The constitutive equations for a viscous fluid

The constitutive equations for a viscous fluid prescribe the relation between the rate of strain tensor $\varepsilon_{\alpha\beta}$ and the *compressive* stress tensor $p_{\alpha\beta}(= -\sigma_{\alpha\beta}$ in the notation of § 6.7). As in the case of the isotropic elastic solid there is nothing to be gained by allowing the symmetric tensor $p_{\alpha\beta}$ to be a linear function of the velocity–derivative tensor

$u_{\alpha,\beta}$. We must, however, allow for a term in the stress independent of the rate of strain, i.e. a stress which can exist in a fluid in uniform motion, or at rest. The simplest constitutive equations are therefore of the form

$$p_{\alpha\beta} = c_{\alpha\beta\rho\sigma}\varepsilon_{\rho\sigma} + q_{\alpha\beta} \tag{6.8.1}$$

and, as the preceding paragraph, we can conclude at once that this relation must be of the form,

$$p_{\alpha\beta} = A\delta_{\alpha\beta}\theta + 2B\varepsilon_{\alpha\beta} + C\delta_{\alpha\beta}, \tag{6.8.2}$$

where
$$\theta = \varepsilon_{\alpha\alpha} = \frac{\partial u_1}{\partial x_1} + \frac{\partial u_2}{\partial x_2} + \frac{\partial u_3}{\partial x_3}. \tag{6.8.3}$$

The rate of strain tensor $\varepsilon_{\alpha\beta} = \frac{1}{2}(u_{\alpha,\beta} + u_{\beta,\alpha})$ can be analysed into two parts:

(1) the dilatational rate of strain $\frac{1}{3}\theta\delta_{\alpha\beta}$, and

(2) the non-dilatational part, $\varepsilon_{\alpha\beta} - \frac{1}{3}\theta\delta_{\alpha\beta}$,

which is equivalent to three pure shearing motions (example 2.4, Chapter 2). Hence the constitutive equation can be rewritten in the form

$$p_{\alpha\beta} = + p\delta_{\alpha\beta} - 2\mu(\varepsilon_{\alpha\beta} - \tfrac{1}{3}\theta\delta_{\alpha\beta}) - \mu_v\theta\delta_{\alpha\beta}, \tag{6.8.4}$$

where $\mu = - B$ is the ordinary coefficient of (shear) viscosity and $\mu_v = - A - \frac{2}{3}B$ is the coefficient of 'bulk' viscosity.

Exercises: (1) A set of tensors $T_{\alpha\beta}$ is said to be invariant under the rotation group if

$$R_{a\alpha}T_{\alpha\beta}R'_{\beta b}$$

is a linear function of the tensors of the given set for all rotations R. Show that each of the following sets is invariant:

(*a*) antisymmetric tensors, ($T_{\alpha\beta} = - T_{\beta\alpha}$),

(*b*) symmetric tensors, ($T_{\alpha\beta} = T_{\beta\alpha}$),

(*c*) multiples of the unit tensor, ($T_{\alpha\beta} = c\delta_{\alpha\beta}$),

(*d*) 'deviators', i.e. tensors with zero trace ($T_{\alpha\alpha} = 0$).

REFERENCES

H. WEYL, *The Classical Groups.* Princeton, 1939 (Chapter II)

J. A. SCHOUTEN, *Tensor Analysis for Physicists.* Oxford, 1951 (p. 156)

CHAPTER VII

Spinors

7.1 Introduction

The classical theory of tensors, as we have introduced it in the preceding chapters, appears to provide a complete algebraical structure, inasmuch as the only objects required in the theory are tensors (scalars being regarded as tensors of rank zero and vectors as tensors of rank unity). However, in 1913 Elie Cartan made the remarkable discovery that there exist geometrical objects called 'spinors' which are even simpler than vectors, in the sense that any vector can be expressed in terms of spinors in a similar fashion to the way in which any second rank tensor can be expressed in terms of vectors. These spinors describe what may be called the *micro-structure* of tensors. The theory of spinors simplifies and extends the theory of tensors and provides the appropriate mathematical apparatus for the quantum theory of particles with intrinsic angular momentum.

The theory of spinors can also be extended to tensor analysis in the space–time manifold of special relativity, and some account of this development is given in the examples at the end of this chapter.

There are two ways of approaching the theory of spinors – either by way of isotropic vectors or by way of Clifford algebra as exemplified by Pauli matrices and Dirac matrices.

7.2 Isotropic vectors

Although there are no vectors which are invariant under rotations in three dimensional space, any prescribed rotation R does possess a set of eigenvectors, one of which is real and lies along the axis of rotation, while the others form a pair of conjugate complex vectors. A complex eigenvector of a rotation is commonly called an 'isotropic vector' – indeed this name was in use before tensor analysis was invented, and before the currency of the special definition of Chapter VI.

The eigenvectors of a rotation are easily discovered by choosing a basis such that the direction Px_3 lies along the axis of rotation. A vector (x_α) is then transformed into the vector $(\bar{x}_\alpha)$ where

$$\begin{aligned} \bar{x}_1 &= x_1 \cos\theta + x_2 \sin\theta, \\ \bar{x}_2 &= -x_1 \sin\theta + x_2 \cos\theta, \\ \bar{x}_3 &= x_3 \end{aligned}$$

θ being the angle of rotation. An eigenvector (x_α) and the associated eigenvalue λ satisfy the equation

$$\bar{x}_\alpha = \lambda x_\alpha,$$

whence we easily find the components of the eigenvectors to be

$$\begin{array}{ll} (0, \quad 0, 1) & \text{for } \lambda = 1, \\ (1, \quad i, 0) & \text{for } \lambda = e^{i\theta}, \\ (1, -i, 0) & \text{for } \lambda = e^{-i\theta}. \end{array} \tag{7.2.1}$$

The conjugate complex eigenvectors $(1, \pm i, 0)$ are called isotropic vectors. Clearly each has zero length.

This result is, of course, true whatever basis is employed. Hence, if $(s_\alpha \pm it_\alpha)$ is a pair of conjugate isotropic vectors, then

$$(s_\alpha \pm it_\alpha).(s_\alpha \pm it_\alpha) = 0, \tag{7.2.2}$$

whence

$$(s_\alpha s_\alpha) = (t_\alpha t_\alpha), \tag{7.2.3}$$

and

$$(s_\alpha t_\alpha) = 0. \tag{7.2.4}$$

An isotropic vector is therefore self-orthogonal, and its real and imaginary parts, s_α and t_α, are orthogonal vectors of equal length.

7.3 The isotropic parameter

An isotropic vector (v_1, v_2, v_3) satisfies the equation (7.2.2)

$$v_1^2 + v_2^2 + v_3^2 = 0, \tag{7.3.1}$$

and hence lies along a (complex) rectilinear generator of the cone

$$x_1^2 + x_2^2 + x_3^2 = 0.$$

The components of an isotropic vector can be expressed in a convenient parametric form by writing

$$\frac{v_1 + iv_2}{v_3} = \xi$$

$$\frac{v_1 - iv_2}{v_3} = -\,\xi^{-1}, \tag{7.3.2}$$

where ξ is the (complex) 'isotropic parameter'.

These equations (7.3.2) are in fact the equations of two planes which intersect in the line of the isotropic vector, i.e. they are the equations of any rectilinear generator of the cone $x_1^2 + x_2^2 + {}_3^2 = 0$.

Now the equation (7.3.1), which defines an isotropic vector as a vector of zero length, is obviously invariant under any rotation about the origin. Hence a rotation must transform an isotropic vector (v_1, v_2, v_3) into another isotropic vector $(\bar{v}_1, \bar{v}_2, \bar{v}_3)$. The isotropic parameter ξ of the first vector must therefore be transformed into the isotropic parameter $\bar{\xi}$ of the second vector. We proceed to find the functional relation between $\bar{\xi}$ and ξ.

The direct calculation of $\bar{\xi}$ in terms of ξ for a general rotation R is curiously difficult and it is much easier to carry out the calculation in two stages by expressing R as the product of two reflexions A and B, as in example 10 of Chapter 1.

Under a reflexion A in the plane

$$a_\alpha x_\alpha = 0 \quad (a_\alpha a_\alpha) = 1$$

the vector v_α is transformed into the vector

$$\bar{v}_\alpha = v_\alpha - 2a_\alpha(a_\beta v_\beta).$$

A short calculation then shows that the isotropic parameter of $\bar{v}_\alpha$ is

$$\bar{\xi} = \frac{-\,a_3\xi + (a_1 + ia_2)}{a_3 + (a_1 - ia_2)\xi}. \tag{7.3.3}$$

This relation between ξ and $\bar{\xi}$ is of the form

$$\bar{\xi} = \frac{\alpha\xi + \beta}{\beta^*\xi - \alpha^*}, \tag{7.3.4}$$

where $$\alpha = -a_3 = \alpha^*,$$
$$\beta = a_1 + ia_2, \quad \beta^* = a_1 - ia_2,$$
and $$\alpha\alpha^* + \beta\beta^* = a_1^2 + a_2^2 + a_3^2 = 1. \tag{7.3.5}$$

In the second reflexion we shall find that

$$\bar{\bar{\xi}} = \frac{\gamma\bar{\xi} + \delta}{\delta^*\bar{\xi} - \gamma^*}$$
$$= \frac{\rho\xi + \sigma}{-\sigma^*\xi + \rho^*} \tag{7.3.6}$$

where $$\rho = \gamma\alpha + \delta\beta^*.$$
$$\sigma = \gamma\beta - \delta\alpha^*,$$

This relation gives the transformation of the isotropic parameter ξ under the rotation $R = BA$. Since

$$\rho\rho^* + \sigma\sigma^* = (\alpha\alpha^* + \beta\beta^*)(\gamma\gamma^* + \delta\delta^*) = 1, \tag{7.3.7}$$

the transformation is of the type known as 'unimodular'.

Since any rotation R can be expressed as the product of a pair of reflexions, we have proved that with each rotation R there is associated a certain unimodular transformation of the isotropic parameters. These unimodular transformations therefore provide a 'representation' of the reflexion group.

7.4 Spinors

If we replace the isotropic parameter ξ by the ratio ξ_1/ξ_0 of a pair of complex numbers, then the transformation equations for a reflexion can be expressed in the form

$$\begin{aligned} \bar{\xi}_0 &= -\alpha^*\xi_0 + \beta^*\xi_1, \\ \bar{\xi}_1 &= \beta\xi_0 + \alpha\xi_1. \end{aligned} \tag{7.4.1}$$

and the transformation equations for a rotation in the form

$$\begin{aligned} \bar{\xi}_0 &= \rho^*\xi_0 - \sigma^*\xi_1 \\ \bar{\xi}_1 &= \sigma\xi_0 + \rho\xi_1. \end{aligned}$$

A spinor is defined to be an ordered pair of complex numbers (ξ_0, ξ_1) which transform according to these equations under the reflexion A and the rotation R, the coefficients being determined by (7.3.5) and the corresponding equations for ρ and σ. We shall often omit the suffixes and speak of the spinor $(\xi) = (\xi_0, \xi_1)$.

To show the relation between a spinor and the isotropic vector with which it is associated, we now rewrite equations (7.3.1) in the form

$$\begin{aligned} \xi_0 v_3 + \xi_1(v_1 - iv_2) &= 0 \\ \xi_0(v_1 + iv_2) - \xi_1 v_3 &= 0. \end{aligned} \tag{7.4.2}$$

The matrix of the coefficients of ξ_0, ξ_1 is

$$\left\| \begin{matrix} v_3 & v_1 - iv_2 \\ v_1 + iv_2 & -v_3 \end{matrix} \right\|$$

and this is equal to

$$v_1\sigma_1 + v_2\sigma_2 + v_3\sigma_3$$

where $\sigma_1 = \left\| \begin{matrix} 0 & 1 \\ 1 & 0 \end{matrix} \right\| \quad \sigma_2 = \left\| \begin{matrix} 0 & -i \\ i & 0 \end{matrix} \right\| \quad \sigma_3 = \left\| \begin{matrix} 1 & 0 \\ 0 & -1 \end{matrix} \right\| \qquad (7.4.3)$

These are the famous spin matrices introduced by Pauli into classical (i.e. non-relativistic) quantum theory.

It is convenient to notice at once the relations

$$\left. \begin{aligned} \sigma_1^2 = I, \quad \sigma_2^2 &= I, \quad \sigma_3^2 = I, \\ \sigma_2\sigma_3 = i\sigma_1 &= -\sigma_3\sigma_2, \\ \sigma_3\sigma_1 = i\sigma_2 &= -\sigma_1\sigma_3, \\ \sigma_1\sigma_2 = i\sigma_3 &= -\sigma_2\sigma_1, \end{aligned} \right\} \tag{7.4.4}$$

(where I is the unit matrix

$$I = \left\| \begin{matrix} 1 & 0 \\ 0 & 1 \end{matrix} \right\|) \tag{7.4.5}$$

From these relations we can deduce at once the equation

$$(x_1\sigma_1 + x_2\sigma_2 + x_3\,\sigma_3)^2 = (x_1^2 + x_2^2 + x_3^2)I. \tag{7.3.6}$$

The matrix associated with the reflexion A is by (7.4.1) and (7.4.3) and (7.3.3)

$$\begin{Vmatrix} -\alpha^* & \beta^* \\ \beta & \alpha \end{Vmatrix} = \begin{Vmatrix} a_3 & a_1 - ia_2 \\ a_1 + ia_2 & -a_3 \end{Vmatrix}$$
$$= a_1\sigma_1 + a_2\sigma_2 + a_3\sigma_3.$$

Hence the spinor $\xi = (\xi_0, \xi_1)$ is transformed by the reflexion A into the spinor $\bar{\xi} = (\bar{\xi}_0, \bar{\xi}_1)$ where (regarding ξ as a row matrix)

$$\bar{\xi} = (a_1\sigma_1 + a_2\sigma_2 + a_3\sigma_3)\xi. \tag{7.4.7}$$

Under a second reflexion B, $\bar{\xi}$ becomes $\bar{\bar{\xi}}$ where

$$\bar{\bar{\xi}} = (b_1\sigma_1 + b_2\sigma_2 + b_3\sigma_3)\bar{\xi}.$$

Hence under the rotation $R = BA$, the spinor ξ becomes the spinor $\bar{\bar{\xi}}$ where

$$\begin{aligned} \bar{\bar{\xi}} &= (b_1\sigma_1 + b_2\sigma_2 + b_3\sigma_3)(a_1\sigma_1 + a_2\sigma_2 + a_3\sigma_3)\xi \\ &= (\rho + i\lambda\sigma_1 + i\mu\sigma_2 + i\nu\sigma_3)\xi \end{aligned} \tag{7.4.8}$$

where
$$\begin{aligned} \rho &= a_1b_1 + a_2b_2 + a_3b_3 = \cos\tfrac{1}{2}\theta, \\ \lambda &= b_2a_3 - b_3a_1 = c_1 \sin\tfrac{1}{2}\theta \\ \mu &= b_3a_1 - b_1a_3 = c_2 \sin\tfrac{1}{2}\theta \\ \nu &= b_1a_2 - b_2a_1 = c_3 \sin\tfrac{1}{2}\theta \end{aligned}$$

where (c_1, c_2, c_3) are the direction cosines of the axis of rotation and θ is the angle of rotation. ρ, λ, μ, ν are the parameters introduced by Euler, Olinde and Rodrigues to express a general rotation.

It is apparent from the preceding formulae that the introduction of the spin matrices considerably simplifies spinor formulae and facilitates spinor calculations.

7.5 Spinors and vectors

The close association of a spinor with an isotropic vector shows that a spinor is undoubtedly a geometric object. To show that these geometric objects are, in some sense, more fundamental than vectors, we shall specialize the representation of a spinor in terms of two homogeneous parameters (ξ_0, ξ_1) by effecting the 'normalization'

$$v_3 = -2\xi_0\xi_1.$$

Equations (7.4.2) then imply that

$$v_1 + iv_2 = -2\xi_1^2, \quad v_1 = \xi_0^2 - \xi_1^2,$$
$$v_1 - iv_2 = +2\xi_0^2, \quad v_2 = i(\xi_0^2 + \xi_1^2). \tag{7.5.1}$$

We can now show that any vector can be expressed in terms of a spinor.

Let $s + it = (v_1, v_2, v_3)$ and $s - it = (v_1^*, v_2^*, v_3^*)$ be the isotropic vectors which are eigenvectors of a rotation R about the axis defined by the real vector (u_1, u_2, u_3).

Consider the vector product

$$[s + it, s - it] = 2i[ts] = -2i[st].$$

Direct calculation shows that the components of this vector product are

$$-2i(\xi_0\xi_0^* + \xi_1\xi_1^*).(w_1, w_2, w_3),$$

where

$$w_\alpha = \xi^*\sigma_\alpha\xi$$
$$= \xi_p^*\sigma_{\alpha(pq)}\xi_q, \tag{7.5.2}$$

$\sigma_{\alpha(pq)}$ being the matrix element of σ_α in row p and column q $(p, q = 0, 1)$.

Now by (7.2.3) the scalar product

$$(s + it\ s - it) = -2i(st),$$

which by direct calculation equals

$$2(\xi_0\xi_0^* + \xi_1\xi_1^*)^2.$$

The identity

$$(v_1v_1^* + \ldots)^2 + (v_2v_3^* - v_3v_2^*)^2 + \ldots$$
$$= (v_1^2 + v_2^2 + v_3^2)((v_1^{*2} + v_2^{*2} + v_3^{*2}),$$

then shows that

$$w_1^2 + w_2^2 + w_3^2 = (\xi_0\xi_0^* + \xi_1\xi_1^*)^2. \tag{7.5.3}$$

Hence the vector with components

$$w_\alpha = \xi^*\sigma_\alpha\xi$$

has a magnitude $\sqrt{w_\alpha w_\alpha} = \xi_0\xi_0^* + \xi_1\xi_1^*$.

7.6 The Clifford algebra

An alternative approach to the theory of spinors is suggested by Dirac's construction of the relativistic wave equation in quantum theory. The three-dimensional analogue of his procedure is to replace the fundamental quadratic form $(x_\alpha x_\alpha)$ for the square of the length of a vector by a linear form

$$(x_\alpha \sigma_\alpha),$$

such that

$$(x_\alpha \sigma_\alpha)^2 = (x_\alpha x_\alpha) I, \tag{7.6.1}$$

where the symbols σ_1, σ_2, σ_3 stand for elements of a certain non-commutative algebra with a unit I. We easily find that the elements σ_α must satisfy the conditions

$$\begin{aligned} \sigma_1{}^2 = \sigma_2{}^2 = \sigma_3{}^2 &= I, \\ \sigma_2\sigma_3 + \sigma_3\sigma_2 &= 0, \\ \sigma_3\sigma_1 + \sigma_1\sigma_3 &= 0, \\ \sigma_1\sigma_2 + \sigma_2\sigma_1 &= 0. \end{aligned} \tag{7.6.2}$$

To relate these conditions to the more restrictive conditions on the Pauli matrices (7.4.4) we write

$$\sigma_2\sigma_3 = i\rho_1, \quad \sigma_3\sigma_1 = i\rho_2, \quad \sigma_1\sigma_2 = i\rho_3,$$

and

$$\sigma_1\sigma_2\sigma_3 = \tau.$$

Then

$$\rho_2\rho_3 = -\sigma_3\sigma_1\sigma_1\sigma_2 = \sigma_2\sigma_3 = i\rho_1, \text{ etc.},$$

and

$$\rho_1{}^2 = -\sigma_2\sigma_3\sigma_2\sigma_3 = I,$$

so that ρ_1, ρ_2, ρ_3 satisfy the conditions for spin matrices.

Moreover τ commutes with ρ_1, ρ_2, ρ_3 and

$$\tau^2 = -I.$$

Hence the simplest representation of the Clifford algebra in three dimensions is obtained by taking

$$\tau = iI,$$

when

$$\rho_\alpha = -i\sigma_\alpha\tau = \sigma_\alpha.$$

We thus obtain the equations (7.4.4) for the Pauli matrices – the simplest example of the 'Clifford algebras' obtained by linearizing the quadratic form $(x_\alpha x_\alpha)$ in a space of n-dimensions.

A reference to equations (7.4.3) shows that *any* square matrix of two rows and two columns

$$\begin{Vmatrix} y_{11} & y_{12} \\ y_{21} & y_{22} \end{Vmatrix}$$

can be expressed in the form

$$x_0 I + x_1\sigma_1 + x_2\sigma_2 + x_3\sigma_3,$$

when
$$\begin{aligned} x_0 &= \tfrac{1}{2}(y_{11} + y_{22}), \\ x_3 &= \tfrac{1}{2}(y_{11} - y_{22}), \\ x_1 &= \tfrac{1}{2}(y_{12} + y_{21}), \\ ix_2 &= \tfrac{1}{2}(y_{21} - y_{12}). \end{aligned}$$

The coefficients x_0, x_1, x_2, x_3, are real if the matrix is Hermitian, i.e. if $y_{\alpha\beta}$ and $y_{\beta\alpha}$ are conjugate complex numbers for all α and β.

The Clifford algebra is 'complete' in the sense that any matrix M which commutes with σ_1, σ_2, σ_3 must be a numerical multiple of I. To prove this we note that M must commute with any Hermitian matrix of two rows and columns, and therefore must commute with the matrix

$$E_{\alpha\beta}$$

whose j, k element is $\delta_{j\alpha}\delta_{k\beta}$. Hence

$$M_{ij}E_{jk} = E_{ij}M_{jk},$$

i.e.
$$M_{i\alpha}\delta_{k\beta} = \delta_{i\alpha}M_{\beta k},$$

and
$$M_{i\alpha} = \delta_{i\alpha}M_{\beta\beta} \quad \text{(not summed with respect to } \beta\text{)},$$

therefore
$$M_{11} = M_{22} \quad \text{and} \quad M_{12} = 0, \quad M_{21} = 0,$$

so that M is a numerical multiple of I.

7.7 The inner automorphisms of the Clifford algebra

From any set of matrices, (σ_α) satisfying the conditions (7.4.4) we can obtain a second set (ρ_α) by means of the equation

$$\rho_\alpha = R_{\alpha j}\sigma_j \tag{7.7.1}$$

where R is the matrix of a rotation (§ 1.6). This follows at once from the equations

$$\begin{aligned} \rho_\alpha\rho_\beta + \rho_\beta\rho_\alpha &= R_{\alpha j}\sigma_j\sigma_k R'_{k\beta} + R_{\beta j}\sigma_j\sigma_k R'_{k\alpha} \\ &= R_{\alpha j}\sigma_j\sigma_k R'_{k\beta} + R_{\beta k}\sigma_k\sigma_j R'_{j\alpha} \\ &= R_{\alpha j}(\sigma_j\sigma_k + \sigma_k\sigma_j)R'_{k\beta} \\ &= 2\delta_{jk}. \end{aligned}$$

Hence any matrix

$$x_0 I + x_\alpha \sigma_\alpha = X \tag{7.7.2}$$

is transformed into a matrix

$$x_0 I + x_\alpha \rho_\alpha = x_0 + x_\alpha R_{\alpha j} \sigma_j = Y, \text{ say.}$$

In technical language the rotation R 'maps' the ring of elements $x_0 I + x_\alpha \sigma_\alpha$ on to itself. The mapping is a one–one relation which preserves the structure of the original ring, i.e. if $X_1 \to Y_1$ and $X_2 \to Y_2$, then

$$a_1 X_1 + a_2 X_2 \to a_1 Y_1 + a_2 Y_2$$

and

$$X_1 X_2 \to Y_1 Y_2,$$

a_1 and a_2 being any complex numbers.

The mapping induced in the ring $x_0 I + x_\alpha \sigma_\alpha$ by a rotation R is therefore an automorphism. The central theorem in this presentation of spinor theory is that such an automorphism can be expressed in the form

$$\rho_\alpha = S \sigma_\alpha S', \tag{7.7.3}$$

where S is a Hermitian matrix with transpose S', such that

$$S'S = I. \tag{7.7.4}$$

Moreover S is uniquely determined apart from a factor ± 1. The following simple proof (given by the author in 1930) gives the form of the matrix S explicitly.

We write

$$\begin{aligned} S &= c(I + \rho_\alpha \sigma_\alpha), \\ S' &= c(I + \sigma_\alpha \rho_\alpha), \end{aligned} \tag{7.7.5}$$

where c is a real number. Then it is easily verified that

$$\begin{aligned} \rho_\alpha S &= S\sigma_\alpha, \quad S'\rho_\alpha = \sigma_\alpha S' \\ \text{and} \qquad \sigma_\alpha S'S &= S'\rho_\alpha S = S'S\sigma_\alpha, \end{aligned} \tag{7.7.6}$$

and $S'S$ commutes with $\sigma_1, \sigma_2, \sigma_3$. This means (as we have proved in 7.6) that $S'S$ is a numerical multiple k of I.
Clearly k cannot be zero, for

$$S'S\sigma_\alpha S' = S'\rho_\alpha \neq 0,$$

and so we can choose c so that $S'S = I$. Hence $\rho_\alpha = S\sigma_\alpha S'$.

Moreover the matrix S is unique, apart from a factor ± 1. For, if there were another matrix T such that

$$\rho_\alpha T = T\sigma_\alpha, \quad T'\rho_\alpha = \sigma_\alpha T',$$

then
$$S'T\sigma_\alpha = S'\rho_\alpha T = \sigma_\alpha S'T,$$

i.e. $S'T$ commutes with σ_1, σ_2, σ_3. Therefore

$$S'T = kI$$

and
$$T = kS.$$

Finally
$$T'T = k^2 S'S = k^2 I,$$

whence
$$T = \pm S.$$

7.8 The spinor manifold

The algebraic investigation of the preceding section shows that with any rotation $R_{\alpha j}$ which carries the spin matrices σ_α into another set of spin matrices ρ_α, so that

$$\rho_\alpha = R_{\alpha j}\sigma_j,$$

we can associate a matrix

$$S = c(I + \rho_\alpha \sigma_\alpha),$$

such that
$$S'S = I,$$

and uniquely determined by R apart from a factor ± 1.

We now propose to provide the matrices S with 'vectors' ξ on which they can operate. Since S has only two rows and two columns, these 'vectors' will have only two components, say ξ_0 and ξ_1. These 'vectors' are in fact the spinors introduced in § 7.4.

The justification for regarding spinors as geometric objects is that they can be associated with the tensor bases in such a way that the spinor transformation from a base PX_α to a base PY_α is determined by the rotation R which relates these two bases. In fact, if

$$\rho_\alpha = R_{\alpha j}\sigma_j = S\sigma_\alpha S',$$

the spinor transformation is taken to be

$$\bar{\xi} = S'\xi,$$

and is therefore determined up to a factor ± 1.

The matrices S provide a 'representation' of the rotation group which is not equivalent to the representations obtained by considering the transformation of any vector or tensor.

Vectors now appear as entities expressible in terms of spinors by means of the relation

$$u_\alpha = \xi^* \sigma_\alpha \xi$$
$$= \xi_j^* \sigma_{\alpha,jk} \xi_k \quad (j, k = 0, 1),$$

$\sigma_{\alpha,jk}$ being matrix elements of σ_α. To verify the transformation law of the vector u under a rotation R we note that from (7.7.1) and (7.7.6) it follows that

$$\rho_{\alpha,mn} = R_{\alpha\beta} \sigma_{\beta,mn}$$
$$= S_{mj} \sigma_{\alpha,jk} S'_{kn}.$$

Hence, if $\quad \bar{\xi}_n = S'_{nk} \xi_k$

then $\quad \bar{\xi}_m^* = \xi_j^* S_{jm}$

since S is Hermitian. Therefore

$$\bar{\xi}_m^* \sigma_{\alpha\, mn} \bar{\xi}_n = \xi_j^* S_{jm} \sigma_{\alpha,mn} S'_{nk} \xi_k$$
$$= \xi_j^* \rho_{\alpha,jk} \xi_k$$
$$= R_{\alpha\beta} \xi_j^* \sigma_{\beta,jk} \xi_k.$$

Thus the spinor transformation,

$$\xi \to \bar{\xi} = S' \xi$$

induces the vector transformation,

$$u \to Ru.$$

Exercises: (1) If R is a rotation about the axis PX_3, then

$$\rho_1 = \sigma_1 \cos\theta + \sigma_2 \sin\theta,$$
$$\rho_2 = -\sigma_1 \sin\theta + \sigma_2 \cos\theta,$$
$$\rho_3 = \sigma_3,$$

and $\quad S = 2c\{(1 + \cos\theta)I - i\sigma_3 \sin\theta\},$

$$SS' = 16c^2 I \cos^2 \tfrac{1}{2}\theta,$$

where $\quad c = \pm \tfrac{1}{4} \sec \tfrac{1}{2}\theta, \quad$ and $\quad S = \pm (I \cos \tfrac{1}{2}\theta - i\sigma_3 \sin \tfrac{1}{2}\theta).$

(2) In the special theory of relativity the fundamental quadratic form is

$$x_\alpha x_\alpha - c^2 t^2 = x_a x_a$$

where $x_0 = ict$ and $a = 0, 1, 2, 3$. If L is a transformation of the orthogonal group which leaves $x_a x_a$ invariant (i.e. a Lorentz transformation), then

$$L'L = LL' = U,$$

U being the unit matrix of four rows and columns. The eigenvectors of L which correspond to an eigenvalue different from ± 1 are isotropic vectors u_α such that $u_a u_a = 0$. The components of an isotropic vector u_a can be expressed in terms of a spinor ξ_0, ξ_1 as

$$\begin{aligned} u_1 &= \xi_0{}^*\xi_1 + \xi_0\xi_1{}^*, \\ iu_2 &= \xi_0{}^*\xi_1 - \xi_0\xi_1{}^*, \\ u_3 &= \xi_0{}^*\xi_0 - \xi_1{}^*\xi_1, \\ -iu_0 &= \xi_0{}^*\xi_0 + \xi_1{}^*\xi_1. \end{aligned}$$

(3) The Clifford algebra for the quadratic form $(x_a x_a)$ is based on four elements α_a (the Dirac matrices) such that

$$(\alpha_a x_a)^2 = x_a x_a,$$

i.e.
$$\alpha_a\alpha_b + \alpha_b\alpha_a = 2\delta_{ab}.$$

These elements generate an algebra isomorphic with the algebra of Hermitian matrices of four rows and four columns. Any Lorentz transformation L induces an automorphism,

$$\beta_a = L_{ab}\alpha_b$$

which can be expressed in the form

$$\beta_a = S\alpha_a S'$$

where
$$S = c \,\Sigma\, \beta_a\alpha_a,$$

the sum being taken over the 32 operators in the *group* generated by the α_a's and the corresponding β_a's. c can be chosen so that

$$SS' = I.$$

REFERENCES

P. A. M. DIRAC, *Proc. Roy. Soc.* A, **117**, 614 (1928)
G. TEMPLE, *Proc. Roy. Soc.* A, **127**, 339 (1930)
R. BRAUER and H. WEYL, *Amer. Journ. of Maths.* **LVII,** 425 (1935)
E. CARTAN, *Leçons sur le Théorie des Spineurs.* Paris (1938)
J. L. SYNGE, *Relativity.* Amsterdam (1956)

CHAPTER VIII

Tensors in Orthogonal Curvilinear Coordinates

8.1 Introduction

The subject of Cartesian tensors should rightly be confined to the strict limits of rectilinear, rectangular coordinate systems, but some of the corresponding results in systems of orthogonal, curvilinear coordinates are so easily obtained and are of such importance in the dynamics of continua that it seems pardonable to stray a short distance outside the Cartesian frame. In this chapter we obtain the well-known formulae for the gradient, divergence and curl of a vector in curvilinear coordinates, and the lesser-known but useful formula for the divergence of a symmetric tensor of the second rank together with the components of the strain tensor.

8.2 Curvilinear orthogonal coordinates

Curvilinear coordinates are introduced into Euclidean space, or a limited region of Euclidean space, by means of a system of three families of surfaces. We shall take the fundamental equations of these surfaces in the parametric form

$$x_\alpha = x_\alpha(y_1, y_2, y_3), \tag{8.2.1}$$

so that the three families of surfaces are given by

$$y_1 = \text{constant}, \quad y_2 = \text{constant} \quad \text{and} \quad y_3 = \text{constant}$$

respectively. The parameters y_1, y_2, y_3 then constitute a system of curvilinear coordinates. The most familiar example is the system of spherical polar coordinates $y_1 = r$, $y_2 = \theta$, $y_3 = \phi$, specified by the equations

$$x = r \sin \theta \cos \phi, \quad y = r \sin \theta \sin \phi, \quad z = r \cos \theta.$$

The parametric equations (8.2.1) specified the *external* properties of a system of curvilinear coordinates, i.e. their relations to the Cartesian coordinate system x_α. The *internal* properties are specified by the expression for the line element, i.e. the distance δs between a pair of neighbouring points with curvilinear coordinates (y_α) and $(y_\alpha + \delta y_\alpha)$. Neglecting terms of higher order

$$\delta s^2 = \delta x_\alpha \, \delta x_\alpha = g_{\rho\sigma} \, dy_\rho \, dy_\sigma, \tag{8.2.2}$$

where $$g_{\rho\sigma} = \frac{\partial x_\alpha}{\partial y_\rho} \cdot \frac{\partial x_\alpha}{\partial y_\sigma}.$$

It is a remarkable fact that in the development of tensor analysis in curvilinear coordinates all the formulae can be expressed in terms of the metric coefficients $g_{\rho\sigma}$. Tensor analysis is thus dependent only on the internal properties of the curvilinear coordinate system.

We shall consider only orthogonal curvilinear coordinates, in which all the results have great simplicity and a special significance in the dynamics of continua. We are therefore dispensed from distinguishing between covariant and contravariant tensors, and do not even need to give their definitions.

In orthogonal curvilinear coordinates the three families of surfaces y_ρ = constant, ($\rho = 1, 2, 3$), are mutually orthogonal at every point. Now the direction cosines of a normal to the surface y_σ = constant are proportional to $\partial y_\sigma / \partial x_\beta$ ($\beta = 1, 2, 3$). Hence the surfaces y_σ = constant and y_ρ = constant are orthogonal if

$$\frac{\partial y_\rho}{\partial x_\alpha} \cdot \frac{\partial y_\sigma}{\partial x_\alpha} = 0 \quad \text{when } \rho \neq \sigma. \tag{8.2.3}$$

Let $$\frac{\partial y_\rho}{\partial x_\alpha} \cdot \frac{\partial y_\sigma}{\partial x_\alpha} = \gamma_{\rho\sigma} \text{ for all } \rho \text{ and } \sigma.$$

Then $$\frac{\partial x_\beta}{\partial y_\rho} \cdot \frac{\partial y_\rho}{\partial x_\alpha} \frac{\partial y_\sigma}{\partial x_\alpha} = \frac{\partial x_\beta}{\partial y_\rho} \gamma_{\rho\sigma}.$$

But by differentiation of the parametric equations we find (8.2.1) that

$$\frac{\partial x_\beta}{\partial y_\rho}\cdot\frac{\partial y_\rho}{\partial x_\alpha}=\delta_{\alpha\beta}.$$

Hence
$$\frac{\partial y_\sigma}{\partial x_\beta}=\delta_{\alpha\beta}\frac{\partial y_\sigma}{\partial x_\alpha}=\frac{\partial x_\beta}{\partial y_\rho}\gamma_{\rho\sigma}$$

and
$$\frac{\partial x_\beta}{\partial y_\tau}\cdot\frac{\partial y_\sigma}{\partial x_\sigma}=g_{\rho\tau}\gamma_{\rho\sigma},$$

i.e.
$$\begin{aligned}\delta_{\sigma\tau}&=g_{\rho\tau}\gamma_{\rho\sigma}\\&=g_{1\tau}\gamma_{1\sigma}+g_{2\tau}\gamma_{2\sigma}+g_{3\tau}\gamma_{3\sigma}\\&=g_{\sigma\tau}\gamma_{\sigma\sigma}\quad(\textit{not}\text{ summed with respect to }\sigma).\end{aligned}$$

Therefore $g_{\sigma\tau}=0$ if $\tau\neq\sigma$. (8.2.4)

The necessity for an occasional suspension of the summation convention is the price we have to pay for the simplicity of the orthogonal curvilinear coordinate system.

The square of the line element can therefore be written as

$$\delta s^2=g_{11}\,\delta y_1{}^2+g_{22}\,\delta y_2{}^2+g_{33}\,\delta y_3{}^2,$$

or, more conveniently as

$$\delta s^2=h_1{}^2\,\delta y_1{}^2+h_2{}^2\,\delta y_2{}^2+h_3{}^2\,\delta y_3{}^2.$$

The element of volume is clearly

$$\delta x_1\,\delta x_2\,\delta x_3=h_1h_2h_3\,\delta y_1\,\delta y_2\,\delta y_3.$$

We also note that

$$\frac{\partial x_\alpha}{\partial y_\sigma}=\frac{\partial x_\beta}{\partial y_\sigma}\delta_{\alpha\beta}=\frac{\partial x_\beta}{\partial y_\sigma}\frac{\partial x_\beta}{\partial y_\rho}\frac{\partial y_\rho}{\partial x_\alpha}=g_{\rho\sigma}\frac{\partial y_\rho}{\partial x_\alpha}=h_\sigma{}^2\frac{\partial y_\sigma}{\partial x_\alpha}$$

(*not* summed with respect to σ). Hence the direction cosines of the surface $y_\sigma=$ constant are also proportional to $\partial x_\alpha/\partial y_\sigma$, ($\alpha=1,2,3$).

8.3 Curvilinear components of tensors

At any point P an orthogonal curvilinear coordinate system (y_α) determines a *basis*, formed by the directions of the normals to the surface $y_\alpha=$ constant ($\alpha=1,2,3$), in the senses in which y_1, y_2, y_3 respectively are increasing. We shall call this the *local* basis at P.

The relation between this basis and the Cartesian basis of the system (x_α) is given by the rotation operator R, whose component $R_{\rho\alpha}$ is equal to the cosine of the angle between the normals to the surfaces y_ρ = constant and x_α = constant. In the rectilinear system the normal to the surface y_ρ = constant has direction cosines proportional to $\partial x_\sigma/\partial y_\rho$ ($\alpha = 1, 2, 3$). But the length of this vector, $\partial x_\alpha/\partial y_\rho$, is

$$\left(\frac{\partial x_1}{\partial y_\rho}\right)^2 + \left(\frac{\partial x_2}{\partial y_\rho}\right)^2 + \left(\frac{\partial x_3}{\partial y_\rho}\right)^2 = g_{\rho\rho} = h_\rho^{\ 2} \quad ,$$

(*not* summed with respect to ρ). Hence the required direction cosines are

$$R_{\rho\alpha} = \frac{\partial x_\alpha}{h_\rho\, \partial y_\rho} \qquad \text{(8.3.1) (n.s.w.t. } \rho)$$

Also since

$$\frac{\partial x_\alpha}{h_\rho\, \partial y_\rho} \cdot \frac{h_\rho\, \partial y_\rho}{\partial x_\beta} = \delta_{\alpha\beta}, \qquad \text{(n.s.w.t. } \rho)$$

we find that in the transposed, or inverse, rotation operator R',

$$R'_{\beta\rho} = \frac{h_\rho\, \partial y_\rho}{\partial x_\beta}. \qquad (8.3.2)$$

The curvilinear components B_ρ of a vector at a point P are defined to be the components of the vector in the local basis at P. Hence if the rectilinear components are A_α, the standard formula (1.6.1) show that

$$B_\rho = R_{\rho\alpha} A_\alpha$$

and

$$A_\beta = R'_{\beta\rho} B_\rho \qquad (8.3.3)$$

8.4 Gradient, divergence and curl in orthogonal curvilinear coordinates

We can obtain some of the more important tensor formulae in orthogonal curvilinear coordinates by direct transformation according to the formulae (8.3.3).

Since the rectilinear components of the gradient of a scalar ϕ are given by

$$A_\alpha = \partial\phi/\partial x_\alpha,$$

it follows at once that the curvilinear components are

$$\begin{aligned} B_\rho &= R_{\rho\alpha}A_\alpha \\ &= \frac{\partial x_\alpha}{hy_\rho\partial} \cdot \frac{\partial\phi}{\partial x_\alpha} = \frac{\partial\phi}{h_\rho \partial y_\rho}. \end{aligned} \quad (8.4.1)$$

To calculate the divergence of a vector B_ρ in curvilinear coordinates we start with the 'flux integral'.

$$F = \iint (A_1\, dx_2\, dx_3 + A_2\, dx_3\, dx_1 + A_3\, dx_1\, dx_2), \quad (8.4.2)$$

taken over a closed surface S. We shall only consider closed surfaces which can be represented by the parametric equations

$$x_\alpha = x_\alpha(u, v),$$

where $x_\alpha(u, v)$ are functions with continuous first derivatives with respect to u and v. Then we can write the flux integral more explicitly as

$$F = \frac{1}{2}\iint \varepsilon_{\alpha\beta\gamma} A_\alpha \frac{\partial(x_\beta, x_\gamma)}{\partial(u, v)}\, du\, dv.$$

Now $\quad \dfrac{\partial(x_\beta, x_\gamma)}{\partial(u, v)} = \dfrac{\partial(x_\beta, x_\gamma)}{\partial(y_\rho, y_\sigma)} \cdot \dfrac{\partial(y_\rho, y_\sigma)}{\partial(u, v)}$

and $\varepsilon_{\alpha\beta\gamma}\dfrac{\partial(x_\beta,x_\gamma)}{\partial(y_\rho,y_\sigma)} = \varepsilon_{\alpha\beta\gamma}h_\rho h_\sigma \begin{vmatrix} R_{\rho\beta} & R_{\rho\gamma} \\ R_{\sigma\beta} & R_{\sigma\gamma} \end{vmatrix}$ from (8.3.1)

$= \varepsilon_{\rho\sigma\tau}h_\rho h_\sigma R_{\tau\alpha}$ by example 7, Chapter 1.

Hence

$$\begin{aligned} F &= \frac{1}{2}\iint \varepsilon_{\rho\sigma\tau}h_\rho h_\sigma B_\tau \frac{\partial(y_\rho, y_\sigma)}{\partial(u, \sigma)}\, du\, dv \\ &= \iint \{h_2h_3B_1\, dy_2\, dy_3 + h_3h_1B_2\, dy_3\, dy_1 \\ &\qquad + h_1h_2B_3\, dy_1\, dy_2\}. \end{aligned} \quad (8.4.3)$$

The divergence of B can now be obtained at once by Green's theorem and the formula,

$$\iiint \operatorname{div} F\, dx_1\, dx_2\, dx_3 = \iiint \operatorname{div} F.\, h_1 h_2 h_3\, dy_1\, dy_2\, dy_3.$$

Therefore

$$\operatorname{div} A = \operatorname{div} B = \frac{1}{h_1 h_2 h_3}\left\{\frac{\partial}{\partial y_1}(h_2 h_3 B_1 + \frac{\partial}{\partial y_2}(h_3 h_1 B_2) + \frac{\partial}{\partial y_3}(h_1 h_2 B_3)\delta\right\}. \tag{8.4.4}$$

Finally to calculate the curl of a vector we form the integral

$$\oint A_\alpha\, dx_\alpha = \oint B_\rho h_\rho\, dy_\rho,$$

taken around any closed curve C with a continuously turning tangent. By Stokes's theorem this line integral is equal to an integral

$$\int \left\{C_1 h_2 h_3\, dy_2\, dy_3 + C_2 h_3 h_1\, dy_3\, dy_1 + C_3 h_1 h_2\, dy_1\, dy_2\right\},$$

taken over any 'smooth' surface S bounded by the curve C, where

$$C_1 h_2 h_3 = \frac{\partial(h_3 B_3)}{\partial y_2} - \frac{\partial(h_2 B_2)}{\partial y_3}, \text{ etc.}$$

Now by comparing the two surface integrals for F (8.4.2) and (8.4.3) we see that if A^*_α and B^*_α denote the components of curl **A** and curl **B** in rectilinear and curvilinear coordinates respectively, then

$$\iint (A^*_1\, dx_2\, dx_3 + A^*_3\, dx_3\, dx_1 + A^*_2\, dx_1\, dx_2)$$

$$= \iint (h_2 h_3 B^*_1\, dy_2\, dy_3 + h_3 h_1 B^*_2\, dy_3\, dy_1 + h_1 h_2 B^*_3\, dy_1\, dy_2).$$

Hence the components of curl **B** in curvilinear coordinates are C_α where

$$C_1 = \frac{1}{h_2 h_3}\left\{\frac{\partial(h_3 B_3)}{\partial y_2} - \frac{\partial(h_2 B_2)}{\partial y_3}\right\},$$

$$C_2 = \frac{1}{h_3 h_1}\left\{\frac{\partial(h_1 B_1)}{\partial y_3} - \frac{\partial(h_3 B_3)}{\partial y_1}\right\}, \tag{8.4.5}$$

$$C_3 = \frac{1}{h_1 h_2}\left\{\frac{\partial(h_2 B_2)}{\partial y_1} - \frac{\partial(h_1 B_1)}{\partial y_2}\right\}.$$

Exercises: (1) Show that for cylindrical polar coordinates (r, ϕ, z) the explicit expressions for grad f, div $\mathbf{B}$ and curl $\mathbf{B}$ are

$$\text{grad}\, f = \left\{\frac{\partial f}{\partial r}, \frac{\partial f}{r\,\partial \phi}, \frac{\partial f}{\partial z}\right\}$$

$$\text{div}\, \mathbf{B} = \frac{\partial(rB_1)}{r\,\partial r} + \frac{\partial B_2}{r\,\partial \phi} + \frac{\partial B_3}{\partial z},$$

$$\text{curl}\, \mathbf{B} = \left\{\frac{\partial B_3}{r\,\partial \phi} - \frac{\partial B_2}{\partial z}, \frac{\partial B_1}{\partial z} - \frac{\partial B_3}{\partial rz}, \frac{\partial(r B_2)}{r\,\partial r} - \frac{\partial B_1}{r\,\partial \phi}\right\}.$$

(2) Show that in spherical polar coordinates (R, θ, ϕ) the explicit expression for grad f, div $\mathbf{B}$ and curl $\mathbf{B}$ are

$$\text{grad}\, f = \frac{\partial f}{\partial R}, \frac{\partial f}{R\partial \phi}, \frac{\partial f}{R \sin\theta \partial \phi},$$

$$\text{div}\, \mathbf{B} = \frac{\partial(R^2 B_1)}{R^2\,\partial R} + \frac{\partial(\sin\theta\, B_2)}{R\sin\theta\,\partial\theta} + \frac{\partial B_3}{R\sin\theta\,\partial\phi},$$

$$\text{curl}\, \mathbf{B} = \left\{\frac{1}{R\sin\theta}\left[\frac{\partial(\sin\theta\, B_3)}{\partial\theta} - \frac{\partial B_2}{\partial\phi}\right], \frac{1}{R\sin\theta}\left[\frac{\partial B_1}{\partial\phi} - \frac{\sin\theta\,\partial(RB_3)}{\partial R}\right], \frac{1}{R}\left[\frac{\partial(RB_2)}{\partial R} - \frac{\partial B_1}{\partial\theta}\right]\right\}.$$

8.5 The strain tensor in orthogonal curvilinear coordinates

The components of the strain tensor (and of the rate of strain tensor) in orthogonal curvilinear coordinates can be easily obtained from the definition given in § 4.4. We consider the displacements from the point P with coordinates y_α, to the point $\bar{P}$ with coordinates $\bar{y}_\alpha$, and from the point Q with coordinates $y_\alpha + \delta y_\alpha$ to the point $\overline{Q}$ with coordinates $\bar{y}_\alpha + \delta\bar{y}_\alpha$.

We have to distinguish between $\eta_\alpha = \bar{y}_\alpha - y_\alpha$, the difference in the coordinates of P and $\bar{P}$, and the corresponding displacement which is given by the vector with components

$$u_\alpha = h_\alpha \eta_\alpha.$$

Now $\qquad |PQ|^2 = \delta s^2 = h_\alpha{}^2\, \delta y_\alpha{}^2$

and $\qquad |\overline{PQ}|^2 = h_\alpha{}^2(y_\rho + \eta_\rho)(\delta y_\alpha + \delta\eta_\alpha)^2,$

where $h_\alpha(y_\rho + \eta_\rho)$ is the value of h_α at the point $\bar{P}$.

Hence

$$|\overline{PQ}|^2 = \left\{h_\alpha{}^2 + 2h_\alpha\frac{\partial h_\alpha}{\partial y_\rho}\eta_\rho\right\}\{\delta y_\alpha{}^2 + 2\delta y_\alpha\delta\eta_\alpha + \delta\eta_\alpha{}^2\},$$

and the direction cosines of $\overrightarrow{PQ}$ are

$$\lambda_\alpha = h_\alpha\, dy_\alpha/ds.$$

Now the invariant I is

$$I = \frac{|\overline{PQ}|^2 - |PQ|^2}{2\,|PQ|^2} = \varepsilon_{\rho\sigma}\lambda_\rho\lambda_\sigma.$$

Hence

$$h_\rho h_\sigma\varepsilon_{\rho\sigma} = \tfrac{1}{2}h_\alpha{}^2\left(\delta_{\alpha\rho} + \frac{\partial\eta_\alpha}{\partial y_\rho}\right)\left(\delta_{\alpha\sigma} + \frac{\partial\eta_\alpha}{\partial y_\sigma}\right) - \tfrac{1}{2}h_\rho{}^2\,\delta_{\rho\sigma}$$
$$+ h_\alpha\frac{\partial h_\alpha}{\partial y_\tau}\eta_\tau\left\{\delta_{\alpha\rho} + \frac{\partial\eta_\alpha}{\partial y_\rho}\right\}\left\{\delta_{\alpha\sigma} + \frac{\partial\eta_\alpha}{\partial y_\sigma}\right\},$$

neglecting terms in $\eta_\sigma\eta_\rho$.

It is convenient to divide the strain tensor into two parts, the part ε^* which is linear in η and its derivatives, and the second ε^+ which contains the remaining terms of higher order.

Then $\varepsilon = \varepsilon^* + \varepsilon^+$,

and
$$\overset{*}{\varepsilon}_{\rho\sigma} = \frac{1}{2}\left\{\frac{h_\rho}{h_\sigma}\cdot\frac{\partial\eta_\rho}{\partial y_\sigma} + \frac{h_\sigma}{h_\rho}\cdot\frac{\partial\eta_\sigma}{\partial y_\rho}\right\} + \sum_\tau \eta_\tau\frac{\partial h_\rho}{h_\rho\,\partial y_\tau}\delta_{\rho\sigma}$$

while
$$\overset{+}{\varepsilon}_{\rho\sigma} = \frac{\partial\eta_\rho}{h_\sigma\,\partial y_\sigma}\cdot\frac{\partial h_\rho}{\partial y_\tau}\eta_\tau + \frac{\partial\eta_\sigma}{h_\rho\,\partial y_\rho}\cdot\frac{\partial h_\sigma}{\partial y_\tau}\eta_\tau$$
$$+ \frac{1}{h_\rho h_\sigma}\left\{\frac{1}{2}h_\alpha{}^2 + h_\alpha\frac{\partial h_\alpha}{\partial y_\tau}\eta_\tau\right\}\frac{\partial\eta_\alpha}{\partial y_\rho}\frac{\partial\eta_\alpha}{\partial y_\sigma}.$$

Usually we need retain only the linear terms and these we can express as functions of the displacement $u_\alpha = h_\alpha \eta_\alpha$. A typical shearing strain ε_{23}^* becomes

$$\varepsilon_{23}^* = \frac{1}{2}\cdot\frac{h_2}{h_3}\cdot\frac{\partial(u_2/h_2)}{\partial y_3} + \frac{1}{2}\cdot\frac{h_3}{h_2}\cdot\frac{\partial(u_3/h_3)}{\partial y_2},$$

and a typical extension or Tensile strain ε_{11}^* becomes

$$\varepsilon_{11}^* = \frac{\partial u_1}{h_1\,\partial y_1} + \frac{u_2}{h_1 h_2}\cdot\frac{\partial h_1}{\partial y_2} + \frac{u_3}{h_1 h_3}\frac{\partial h_1}{\partial y_3}.$$

Examples: (1) Show that in cylindrical polar coordinates

$$(y_1 = r,\ y_2 = \phi,\ y_3 = z),\ (u_1 = u,\ u_2 = v,\ u_3 = w),$$

the components of ε^* are

$$\varepsilon_{11}^* = \frac{\partial u}{\partial r},\quad \varepsilon_{22}^* = \frac{\partial v}{r\,\partial\phi} + \frac{u}{r},\quad \varepsilon_{33}^* = \frac{\partial w}{\partial z},$$

$$\varepsilon_{23}^* = \frac{1}{2}\left(\frac{\partial w}{r\,\partial\phi} + \frac{\partial v}{\partial z}\right),\quad \varepsilon_{31}^* = \frac{1}{2}\left(\frac{\partial u}{\partial z} + \frac{\partial w}{\partial r}\right),$$

$$\varepsilon_{12}^* = \frac{1}{2}\left(\frac{\partial v}{\partial r} - \frac{v}{r} + \frac{\partial u}{r\,\partial\phi}\right).$$

(2) Show that in spherical polar coordinates

$$(y_1 = R,\ y_2 = \theta,\ y_3 = \phi),\ (u_1 = u,\ u_2 = v,\ u_3 = w),$$

the components of ε^* are

$$\varepsilon_{11}^* = \frac{\partial u}{\partial R},\quad \varepsilon_{22}^* = \frac{\partial v}{R\,\partial\theta} + \frac{u}{R},\quad \varepsilon_{33}^* = \frac{\partial w}{R\sin\theta\,\partial\phi} + \frac{v\cot\theta}{R} + \frac{u}{R},$$

$$\varepsilon_{23}^* = \frac{1}{2}\left(\frac{\partial w}{R\,\partial\theta} - \frac{w\cot\theta}{R} + \frac{\partial v}{R\sin\theta\,\partial\phi}\right),$$

$$\varepsilon_{31}^* = \frac{1}{2}\left(\frac{\partial u}{R\sin\theta\,\partial\phi} + \frac{\partial w}{\partial R} - \frac{w}{R}\right),$$

$$\varepsilon_{12}^* = \frac{1}{2}\left(\frac{\partial v}{\partial R} - \frac{v}{R} + \frac{\partial u}{R\,\partial\theta}\right).$$

8.6 The three index symbols

The simple construction of expressions in curvilinear coordinates for the gradient, divergence and curl of a vector is almost a happy accident. In general the deeper analysis of Riemannian geometry is necessary to construct expressions for the rate of change of a vector or tensor in an arbitrary direction. It is, however, just possible to obtain the formula for the divergence of a tensor in curvilinear coordinates without using all the formal apparatus of covariant differentiation. Since this formula is needed in forming the equations of motion of a continuous medium we shall stretch our Cartesian tensor theory far enough to yield this result.

We do, however, need one formula of Riemannian geometry, viz the expressions for the 'three index symbols' introduced by Christoffel. These are

$$\Gamma_{\rho\sigma\tau} \equiv \frac{1}{2}\left\{\frac{\partial g_{\sigma\tau}}{\partial y_\rho} + \frac{\partial g_{\rho\tau}}{\partial y_\sigma} - \frac{\partial g_{\rho\sigma}}{\partial y_\tau}\right\}$$

$$= \frac{\partial^2 x_\alpha}{\partial y_\rho\, \partial y_\sigma} \cdot \frac{\partial x_\alpha}{\partial y_\tau}. \tag{8.6.1}$$

This result is easily proved by differentiating the fundamental identity (8.2.2)

$$g_{\rho\sigma} = \frac{\partial x_\alpha}{\partial y_\rho} \cdot \frac{\partial x_\alpha}{\partial y_\sigma},$$

then we find that

$$\frac{\partial g_{\rho\sigma}}{\partial y_\tau} = \frac{\partial^2 x_\alpha}{\partial y_\rho\, \partial y_\tau} \cdot \frac{\partial x_\alpha}{\partial y_\sigma} + \frac{\partial x_\alpha}{\partial y_\rho} \cdot \frac{\partial^2 x_\alpha}{\partial y_\tau\, \partial y_\sigma}$$

$$= (\rho, \tau)\,(\sigma) + (\rho)\,(\tau, \sigma),$$

in an obvious notation. Hence, by cyclic permutation of ρ, σ, τ we find that

$$\frac{\partial g_{\sigma\tau}}{\partial y_\rho} = (\sigma, \rho)(\tau) + (\sigma)(\rho, \tau),$$

$$\frac{\partial g_{\rho\tau}}{\partial y_\sigma} = (\tau, \sigma)(\rho) + (\tau)\,(\sigma, \rho).$$

Therefore $$\Gamma_{\rho\sigma\tau} \equiv (\sigma, \rho)(\tau) = \frac{\partial^2 x_\alpha}{\partial y_\rho \, \partial y_\sigma} \cdot \frac{\partial x_\alpha}{\partial y_\tau}.$$

8.7 The divergence of the stress tensor in curvilinear coordinates

In studying the dynamics of a continuous medium we need expressions for the 'body forces' per unit volume of a region V which are equivalent to the surface stresses acting on the surface S which bounds V. In rectilinear coordinates the equivalent body forces are easily formed as follows:

A surface stress given by the tensor S_{ab} acting on a surface element dS with direction cosines (λ_a) produces a force with component

$$l_a S_{ab} \lambda_b . dS$$

in the direction given by the unit vector (l_a). Hence the resultant force due to the stresses on a closed surface S is

$$\iint l_a S_{ab} \lambda_b . dS.$$

By Green's theorem this is equal to the integral

$$\iiint \frac{\partial(l_a S_{ab})}{\partial x_b} \, d\tau$$

taken throughout the volume V bounded by S. Therefore the volume density of the equivalent body forces is represented by the vector

$$\partial S_{ab}/\partial x_b, \qquad (8.7.1)$$

which is commonly called the 'divergence' of the tensor S_{ab}.

We now proceed to use the same method to calculate the divergence of the stress tensor in rectangular curvilinear coordinates. Of course the same method applies to any symmetric second rank tensor.

In a system of rectangular curvilinear coordinates (y_α) let the components of the stress tensor be $T_{\alpha\beta}$, and the components of the unit vector (l_a) be (m_α). Let R be the rotation operator which relates tensors in the rectilinear basis (x_α) and the curvilinear basis (y_α).

Then

$$R_{\rho\alpha} = \frac{\partial x_\alpha}{h_\rho \partial y_\rho},$$

$$R'_{\beta\rho} = \frac{h_\rho\, \partial y_\rho}{\partial x_\beta},$$

$$m_\rho = R_{\rho a} l_a, \quad l_b = R'_{b\rho} m_\rho = m_\rho R_{\rho b},$$

$$S_{ab} = R'_{a\alpha} T_{\alpha\beta} R_{\beta b}.$$

Consider as in 8.4 the resultant (in the direction of the unit vector l_a) of the stresses acting on the 'smooth' surface S with parametric equations

$$x_\alpha = x_\alpha(u, v)$$

or

$$y_\alpha = y_\alpha(u, v).$$

It is

$$\frac{1}{2}\iint \varepsilon_{abc} S_{ad} l_d \frac{\partial(x_b, x_c)}{\partial(u, v)}\, du\, dv. \tag{8.7.2}$$

As before

$$\varepsilon_{abc} \frac{\partial(x_b, x_c)}{\partial(u, v)} = \varepsilon_{\rho\sigma\tau} h_\rho h_\sigma R_{\sigma a} \frac{\partial(y_\rho, y_\sigma)}{\partial(u, v)},$$

whence the integrand of (8.7.2) is

$$\tfrac{1}{2}\varepsilon_{\rho\sigma\tau} h_\rho h_\sigma \frac{\partial(y_\rho, y_\sigma)}{(u, v)} R_{\tau a} R'_{a\alpha} T_{\alpha\beta} R_{\beta d} R'_{dc} m_c$$

$$= \tfrac{1}{2}\varepsilon_{\rho\sigma\tau} h_\rho h_\sigma \frac{\partial(y_\rho, y_\sigma)}{\partial(u, v)} T_{\tau\beta} m_\beta$$

$$= \tfrac{1}{2}\varepsilon_{\rho\sigma\tau} \frac{h}{h_\tau} \cdot \frac{\partial(y_\rho, y_\sigma)}{\partial(u, v)} T_{\tau\beta} m_\beta,$$

where $\qquad h = h_1 h_2 h_3.$

The surface integral of this expression is, by Green's theorem, equal to the volume integral of

$$B_\beta m_\beta = \frac{1}{h} \cdot \frac{\partial}{\partial y_\tau}\left\{\frac{h}{h_\tau} T_{\tau\beta} m_\beta\right\} = \frac{\partial}{h\, \partial y_\tau}\left\{\frac{h h_\beta}{h_\tau} T_{\tau\beta} \frac{\partial x_\alpha}{h_\beta^2\, \partial y_\beta} l_\alpha\right\}$$

$$= \frac{1}{h} \cdot \frac{\partial}{\partial y_\tau}\left\{\frac{h h_\beta}{h_\tau} T_{\tau\beta}\right\} \frac{m_\beta}{h_\beta} + \frac{h_\beta}{h_\tau} T_{\tau\beta} \frac{\partial}{\partial y_\tau}\left\{\frac{\partial x_\alpha}{h_\beta^2\, \partial y_\beta}\right\} l_\alpha. \tag{8.7.3}$$

The coefficient of $T_{\tau\beta}$ in (8.7.3) is

$$C_{\beta\tau} = \frac{h_\beta}{h_\tau}\left\{\frac{\partial^2 x_\alpha}{h_\beta^2\,\partial y_\tau\,\partial y_\beta} - \frac{2\partial h_\beta}{h_\beta^3\,\partial y_\tau}\cdot\frac{\partial x_\alpha}{\partial y_\beta}\right\}m_\rho\frac{\partial x_\alpha}{h_\rho\,\partial y_\rho}$$

$$= \frac{\partial^2 x_\alpha}{h_\rho h_\tau h_\beta\,\partial y_\tau\,\partial y_\beta}\cdot\frac{\partial x_\alpha}{\partial y_\rho}m_\rho - \frac{2\partial h_\beta}{h_\tau h_\beta^3\,\partial y_\tau}m_\rho g_{\beta\rho}. \qquad (8.7.4)$$

Now by (8.5.1)

$$\frac{\partial^2 x_\alpha}{\partial y_\tau\,\partial y_\beta}\cdot\frac{\partial x_\alpha}{\partial y_\rho} = \Gamma_{\tau\beta\rho} = \frac{1}{2}\left\{\frac{\partial g_{\tau\rho}}{\partial y_\beta} + \frac{\partial g_{\beta\rho}}{\partial y_\tau} - \frac{\partial g_{\tau\beta}}{\partial y_\rho}\right\}.$$

Hence, changing the dummy suffix β into ρ in the last term, we find that

$$C_{\beta\tau}T_{\tau\beta} = m_\rho\left\{\frac{T_{\rho\beta}\,\partial h_\rho}{h_\rho h_\beta\,\partial y_\beta} + \frac{T_{\tau\rho}\,\partial h_\rho}{h_\tau h_\rho\,\partial y_\tau} - \frac{T_{\beta\beta}\,\partial h_\beta}{h_\rho h_\beta\,\partial y_\rho} - \frac{2T_{\tau\rho}\,\partial h_\rho}{h_\tau h_\rho\,\partial y_\tau}\right\}$$

$$= -\frac{T_{\beta\beta}\,\partial h_\beta}{h_\beta h_\rho\,\partial y_\rho}m_\rho = -\frac{T_{\alpha\alpha}\partial h_\alpha}{h_\beta h_\alpha\,\partial y_\beta}\cdot m_\beta.$$

Therefore

$$B_\beta = \frac{1}{h_\beta h}\cdot\frac{\partial}{\partial y_\tau}\left\{\frac{hh_\beta}{h_\tau}T_{\tau\beta}\right\} - \frac{T_{\alpha\alpha}\,\partial h_\alpha}{h_\beta h_\alpha\,\partial y_\beta},$$

and this is the component of the body force in the direction m_β.

Examples: (It is convenient to denote the components of the stress tensors in the curvilinear coordinates $y_1 = \alpha$, $y_2 = \beta$, $y_3 = \gamma$ by

$$\sigma_{11} = \widehat{\alpha\alpha}, \quad \sigma_{22} = \widehat{\beta\beta}, \quad \sigma_{33} = \widehat{\gamma\gamma},$$

$$\sigma_{23} = \widehat{\beta\gamma}, \quad \sigma_{31} = \widehat{\gamma\alpha}, \quad \sigma_{12} = \widehat{\alpha\beta}).$$

(1) Show that in cylindrical polar coordinates (r, ϕ, z),

$$ds^2 = dr^2 + r^2\,d\phi^2 + dz^2,$$

and the divergence of the stress has components

$$\frac{\partial(\widehat{rr})}{\partial r} + \frac{\partial(\widehat{r\phi})}{r\,\partial\phi} + \frac{\partial(\widehat{rz})}{\partial z} + \frac{\widehat{rr} - \widehat{\phi\phi}}{r},$$

$$\frac{\partial(\widehat{r\phi})}{\partial r} + \frac{\partial)\widehat{\phi\phi})}{r\,\partial\phi} + \frac{\partial(\widehat{\phi z})}{\partial z} + \frac{2\widehat{r\phi}}{r},$$

$$\frac{\partial(\widehat{rz})}{\partial r} + \frac{\partial(\widehat{\phi z})}{r\,\partial\phi} + \frac{\partial(\widehat{zz})}{\partial z} + \frac{\widehat{rz}}{r}.$$

(2) Show that in spherical polar coordinates (R, θ, ϕ),

$$ds^2 = dR^2 + R^2 d\theta^2 + R^2 \sin^2\theta\, d\phi^2,$$

and the divergence of the stress has components

$$\frac{\partial\widehat{RR}}{\partial R} + \frac{\partial\widehat{R\theta}}{R\,\partial\theta} + \frac{\partial\widehat{R\phi}}{R\sin\theta\,\partial\phi} + \frac{1}{R}(2\widehat{RR} - \widehat{\theta\theta} - \widehat{\phi\phi} + \widehat{R\theta}\cot\theta),$$

$$\frac{\partial\widehat{R\theta}}{\partial R} + \frac{\partial\widehat{\theta\theta}}{R\,\partial\theta} + \frac{\partial\widehat{\theta\phi}}{R\sin\theta\,\partial\phi} + \frac{1}{R}\{(\widehat{\theta\theta} - \widehat{\phi\phi})\cot\theta + 3\widehat{R\theta}\},$$

$$\frac{\partial\widehat{R\phi}}{\partial R} + \frac{\partial\widehat{\theta\phi}}{R\,\partial\theta} + \frac{\partial\widehat{\phi\phi}}{R\sin\theta\,\partial\phi} + \frac{1}{R}(3\widehat{R\phi} + 2\widehat{\theta\phi}\cot\theta).$$

Index

Index

A CATALOG OF SELECTED

DOVER BOOKS

IN SCIENCE AND MATHEMATICS

Math–Geometry and Topology

ELEMENTARY CONCEPTS OF TOPOLOGY, Paul Alexandroff. Elegant, intuitive approach to topology from set-theoretic topology to Betti groups; how concepts of topology are useful in math and physics. 25 figures. 57pp. 5⅜ x 8½. 60747-X

COMBINATORIAL TOPOLOGY, P. S. Alexandrov. Clearly written, well-organized, three-part text begins by dealing with certain classic problems without using the formal techniques of homology theory and advances to the central concept, the Betti groups. Numerous detailed examples. 654pp. 5⅜ x 8½. 40179-0

EXPERIMENTS IN TOPOLOGY, Stephen Barr. Classic, lively explanation of one of the byways of mathematics. Klein bottles, Moebius strips, projective planes, map coloring, problem of the Koenigsberg bridges, much more, described with clarity and wit. 43 figures. 210pp. 5⅜ x 8½. 25933-1

CONFORMAL MAPPING ON RIEMANN SURFACES, Harvey Cohn. Lucid, insightful book presents ideal coverage of subject. 334 exercises make book perfect for self-study. 55 figures. 352pp. 5⅝ x 8¼. 64025-6

THE GEOMETRY OF RENÉ DESCARTES, René Descartes. The great work founded analytical geometry. Original French text, Descartes's own diagrams, together with definitive Smith-Latham translation. 244pp. 5⅜ x 8½. 60068-8

PRACTICAL CONIC SECTIONS: The Geometric Properties of Ellipses, Parabolas and Hyperbolas, J. W. Downs. This text shows how to create ellipses, parabolas, and hyperbolas. It also presents historical background on their ancient origins and describes the reflective properties and roles of curves in design applications. 1993 ed. 98 figures. xii+100pp. 6½ x 9¼. 42876-1

THE THIRTEEN BOOKS OF EUCLID'S ELEMENTS, translated with introduction and commentary by Thomas L. Heath. Definitive edition. Textual and linguistic notes, mathematical analysis. 2,500 years of critical commentary. Unabridged. 1,414pp. 5⅜ x 8½. Three-vol. set. Vol. I: 60088-2 Vol. II: 60089-0 Vol. III: 60090-4

GEOMETRY OF COMPLEX NUMBERS, Hans Schwerdtfeger. Illuminating, widely praised book on analytic geometry of circles, the Moebius transformation, and two-dimensional non-Euclidean geometries. 200pp. 5⅝ x 8¼. 63830-8

DIFFERENTIAL GEOMETRY, Heinrich W. Guggenheimer. Local differential geometry as an application of advanced calculus and linear algebra. Curvature, transformation groups, surfaces, more. Exercises. 62 figures. 378pp. 5⅜ x 8½. 63433-7

CURVATURE AND HOMOLOGY: Enlarged Edition, Samuel I. Goldberg. Revised edition examines topology of differentiable manifolds; curvature, homology of Riemannian manifolds; compact Lie groups; complex manifolds; curvature, homology of Kaehler manifolds. New Preface. Four new appendixes. 416pp. 5⅜ x 8½. 40207-X

History of Math

THE WORKS OF ARCHIMEDES, Archimedes (T. L. Heath, ed.). Topics include the famous problems of the ratio of the areas of a cylinder and an inscribed sphere; the measurement of a circle; the properties of conoids, spheroids, and spirals; and the quadrature of the parabola. Informative introduction. clxxxvi+326pp; supplement, 52pp. 5⅜ x 8½. 42084-1

A SHORT ACCOUNT OF THE HISTORY OF MATHEMATICS, W. W. Rouse Ball. One of clearest, most authoritative surveys from the Egyptians and Phoenicians through 19th-century figures such as Grassman, Galois, Riemann. Fourth edition. 522pp. 5⅜ x 8½. 20630-0

THE HISTORY OF THE CALCULUS AND ITS CONCEPTUAL DEVELOPMENT, Carl B. Boyer. Origins in antiquity, medieval contributions, work of Newton, Leibniz, rigorous formulation. Treatment is verbal. 346pp. 5⅜ x 8½. 60509-4

THE HISTORICAL ROOTS OF ELEMENTARY MATHEMATICS, Lucas N. H. Bunt, Phillip S. Jones, and Jack D. Bedient. Fundamental underpinnings of modern arithmetic, algebra, geometry, and number systems derived from ancient civilizations. 320pp. 5⅜ x 8½. 25563-8

A HISTORY OF MATHEMATICAL NOTATIONS, Florian Cajori. This classic study notes the first appearance of a mathematical symbol and its origin, the competition it encountered, its spread among writers in different countries, its rise to popularity, its eventual decline or ultimate survival. Original 1929 two-volume edition presented here in one volume. xxviii+820pp. 5⅜ x 8½. 67766-4

GAMES, GODS & GAMBLING: A History of Probability and Statistical Ideas, F. N. David. Episodes from the lives of Galileo, Fermat, Pascal, and others illustrate this fascinating account of the roots of mathematics. Features thought-provoking references to classics, archaeology, biography, poetry. 1962 edition. 304pp. 5⅜ x 8½. (Available in U.S. only.) 40023-9

OF MEN AND NUMBERS: The Story of the Great Mathematicians, Jane Muir. Fascinating accounts of the lives and accomplishments of history's greatest mathematical minds–Pythagoras, Descartes, Euler, Pascal, Cantor, many more. Anecdotal, illuminating. 30 diagrams. Bibliography. 256pp. 5⅜ x 8½. 28973-7

HISTORY OF MATHEMATICS, David E. Smith. Nontechnical survey from ancient Greece and Orient to late 19th century; evolution of arithmetic, geometry, trigonometry, calculating devices, algebra, the calculus. 362 illustrations. 1,355pp. 5⅜ x 8½. Two-vol. set. Vol. I: 20429-4 Vol. II: 20430-8

A CONCISE HISTORY OF MATHEMATICS, Dirk J. Struik. The best brief history of mathematics. Stresses origins and covers every major figure from ancient Near East to 19th century. 41 illustrations. 195pp. 5⅜ x 8½. 60255-9

Mathematics

FUNCTIONAL ANALYSIS (Second Corrected Edition), George Bachman and Lawrence Narici. Excellent treatment of subject geared toward students with background in linear algebra, advanced calculus, physics, and engineering. Text covers introduction to inner-product spaces, normed, metric spaces, and topological spaces; complete orthonormal sets, the Hahn-Banach Theorem and its consequences, and many other related subjects. 1966 ed. 544pp. 6⅛ x 9¼. 40251-7

ASYMPTOTIC EXPANSIONS OF INTEGRALS, Norman Bleistein & Richard A. Handelsman. Best introduction to important field with applications in a variety of scientific disciplines. New preface. Problems. Diagrams. Tables. Bibliography. Index. 448pp. 5⅜ x 8½. 65082-0

VECTOR AND TENSOR ANALYSIS WITH APPLICATIONS, A. I. Borisenko and I. E. Tarapov. Concise introduction. Worked-out problems, solutions, exercises. 257pp. 5⅝ x 8¼. 63833-2

THE ABSOLUTE DIFFERENTIAL CALCULUS (CALCULUS OF TENSORS), Tullio Levi-Civita. Great 20th-century mathematician's classic work on material necessary for mathematical grasp of theory of relativity. 452pp. 5⅝ x 8¼. 63401-9

AN INTRODUCTION TO ORDINARY DIFFERENTIAL EQUATIONS, Earl A. Coddington. A thorough and systematic first course in elementary differential equations for undergraduates in mathematics and science, with many exercises and problems (with answers). Index. 304pp. 5⅜ x 8½. 65942-9

FOURIER SERIES AND ORTHOGONAL FUNCTIONS, Harry F. Davis. An incisive text combining theory and practical example to introduce Fourier series, orthogonal functions and applications of the Fourier method to boundary-value problems. 570 exercises. Answers and notes. 416pp. 5⅜ x 8½. 65973-9

COMPUTABILITY AND UNSOLVABILITY, Martin Davis. Classic graduate-level introduction to theory of computability, usually referred to as theory of recurrent functions. New preface and appendix. 288pp. 5⅜ x 8½. 61471-9

ASYMPTOTIC METHODS IN ANALYSIS, N. G. de Bruijn. An inexpensive, comprehensive guide to asymptotic methods–the pioneering work that teaches by explaining worked examples in detail. Index. 224pp. 5⅜ x 8½ 64221-6

APPLIED COMPLEX VARIABLES, John W. Dettman. Step-by-step coverage of fundamentals of analytic function theory–plus lucid exposition of five important applications: Potential Theory; Ordinary Differential Equations; Fourier Transforms; Laplace Transforms; Asymptotic Expansions. 66 figures. Exercises at chapter ends. 512pp. 5⅜ x 8½. 64670-X

INTRODUCTION TO LINEAR ALGEBRA AND DIFFERENTIAL EQUATIONS, John W. Dettman. Excellent text covers complex numbers, determinants, orthonormal bases, Laplace transforms, much more. Exercises with solutions. Undergraduate level. 416pp. 5⅜ x 8½. 65191-6